Multi-Agent Technologies and Machine Learning

Edited by Igor A. Sheremet

Published in London, United Kingdom

Multi-Agent Technologies and Machine Learning
http://dx.doi.org/10.5772/intechopen.100659
Edited by Igor A. Sheremet

Contributors
Jaslin Shaleem Khan, Malligama Arachchige Udhitha Sudeera Navarathne, Janaka Bandara Ekanayake, Aida Huerta Barrientos, Alejandro Nila Luevano, Evgeniy Zaytsev, Elena Nurmatova, Fahmina Taranum, Maniza Hijab, Sridevi K, Syeda Fouzia Sayeedunissa, Afshan Kaleem, Niraja K.S, Theocharis Kravaris, George A. Vouros, Zhou Shaorui, Cai Ming, Zhuo Xiaopo

First published in London, United Kingdom, 2023 by IntechOpen
IntechOpen is the global imprint of INTECHOPEN LIMITED, registered in England and Wales, registration number: 11086078, 5 Princes Gate Court, London, SW7 2QJ, United Kingdom

British Library Cataloguing-in-Publication Data
A catalogue record for this book is available from the British Library

Additional hard and PDF copies can be obtained from orders@intechopen.com

Multi-Agent Technologies and Machine Learning
Edited by Igor A. Sheremet
p. cm.

This title is part of the Artificial Intelligence Book Series, Volume 14
Topic: Multi-Agent System
Series Editor: Andries Engelbrecht
Topic Editor: Mehmet Emin Aydin

Print ISBN 978-1-80356-443-2
Online ISBN 978-1-80356-444-9
eBook (PDF) ISBN 978-1-80356-445-6
ISSN 2633-1403

IntechOpen Book Series

Artificial Intelligence

Volume 14

Aims and Scope of the Series

Artificial Intelligence (AI) is a rapidly developing multidisciplinary research area that aims to solve increasingly complex problems. In today's highly integrated world, AI promises to become a robust and powerful means for obtaining solutions to previously unsolvable problems. This Series is intended for researchers and students alike interested in this fascinating field and its many applications.

Meet the Series Editor

Andries Engelbrecht received the Masters and Ph.D. degrees in Computer Science from the University of Stellenbosch, South Africa, in 1994 and 1999 respectively. He is currently appointed as the Voigt Chair in Data Science in the Department of Industrial Engineering, with a joint appointment as Professor in the Computer Science Division, Stellenbosch University. Prior to his appointment at Stellenbosch University, he has been at the University of Pretoria, Department of Computer Science (1998-2018), where he was appointed as South Africa Research Chair in Artifical Intelligence (2007-2018), the head of the Department of Computer Science (2008-2017), and Director of the Institute for Big Data and Data Science (2017-2018). In addition to a number of research articles, he has written two books, Computational Intelligence: An Introduction and Fundamentals of Computational Swarm Intelligence.

Meet the Volume Editor

Prof. Dr. Igor A. Sheremet is a full member of the Russian Academy of Sciences (RAS) and a scientific supervisor and chair of the Science-Technological Board of the National Computer Corporation (NCC), Russia. His main areas of research are systems analysis, artificial intelligence, data science, and digital economy. He is a co-chair of the Advanced Mathematical Tools for Data-Driven Applied Systems Analysis task group established by the Committee on Data (CODATA) of the International Science Council (ISC), a vice-chair of the Committee for Systems Analysis of the RAS, and a vice-chair of the RAS Scientific Board for Robotics and Mechatronics. During 2014–2020 he was a member of the Science Advisory Committee (SAC) of the International Institute for Applied Systems Analysis (IIASA). During 2020–2022 he was a chair of the joint CODATA-IIASA working group Big Data and Systems Analysis.

Contents

Preface

Multi-agent technologies (MATs) and machine learning (ML) are among the most advanced directions of modern computer science and its applications due to their ability to reduce the complexity of and time spent on knowledge acquisition and maintenance. They are also able to achieve high-quality solutions to various problems.

MATs, which arose from agent-based programming, are an efficient tool for the creation and evolutionary development of multi-agent systems (MASs), representing the logic of solution of various non-trivial tasks from multiple problem areas as well as the logic of behavior of various objects, operating in complicated and volatile environments. The great flexibility of MATs, achieved due to the representation of a task or a system as an easily modified set of interacting objects named agents, enables their wide application as well as their software and even hardware implementations. Today, many successful applications of MATs in engineering, physics, medicine, energy, economy, banking, social management, ecology, and more demonstrate the great power of this tool.

Machine learning ML as a segment of artificial intelligence (AI) arose primarily from neural networks (NNs), which are efficient and widely used for solving tasks requiring massive parallelism for processing intensive data flows originating primarily from various sensors. The most well-known application of NNs is image processing, which enables online monitoring of wide areas covered by video cameras and/or measurement devices. A great achievement of ML associated with NNs is its implementation in teaching NNs by input–output sets ("training samples"). Thus, a work of a learning person in this case is reduced to the presentation of an a priori accumulated set of training samples to an NN, transferred to a teaching regime. Moreover, there are ML paradigms such as unsupervised learning and reinforcement learning (or learning with feedback) that do not require a person at all or require them in a very limited capacity.

Since some period of mutually isolated research of MATs and ML, it has become clear that there is much promise in combining the advantages of both approaches: the generality and flexibility inherent to MATs with the low-cost knowledge acquisition inherent to ML. There may be different approaches to constructive development and implementation of this basic idea, for example, the creation of MASs with self-learning agents or MASs with adaptive agents set, and thus investigation of ways of integrating MATs and ML is an interesting and promising area of research.

This book consists of six chapters in three sections. Section 1, which includes Chapters 1 and 2, discusses the integration of MATs and ML. Chapter 1 reviews current research in MASs combined with various ML techniques. It analyzes features of MASs from the ML point of view, classifies applications of MASs integrated with ML methods, and presents a density map of applications in manufacturing, commerce, and E-learning. Chapter 2 discusses deep multi-agent reinforcement learning methods for solving problems with MASs consisting of very large numbers of heterogeneous agents. Such methods have significant scalability potential and inter-agent coordination capabilities in large-scale and complex multi-agent settings. The chapter uses air traffic management as a practical background for comparative consideration of the described methods.

Section 2, which includes Chapters 3 and 4, describes the most interesting and refined applications of MATs combined with ML. Chapter 3 is a comparative study of different design approaches to a corporative multi-agent system for optimal scheduling. The authors propose creating a dataset using multiple algorithms with different performance metrics to find the best one. This dataset may be imported into some ML tools for training and predicting, based on the selected performance metrics. The chapter examines three approaches: first come first serve, round robin, and ant colony. The chapter shows that the ant colony algorithm achieves the best results. Chapter 4 describes an architecture of a knowledge-based multi-agent system (KBMAS) demonstrating inter alia capabilities, which open due to the integration of advantages of MATs and ML. It proposes an approach to the development of applied software agents that combines knowledge-based reasoning with neural network models. It also considers the method of reinforcement learning, the system of rules, and the queries to the knowledge base.

Section 3, which includes Chapters 5 and 6, describes advanced results in MATs and ML that may be used as a background for their further integration and extension of an area of their combined theoretical consideration and application. Chapter 5 discusses the multi-agent modeling of a charging station of an electric vehicle. Power engineering applications, such as power system optimization and restoration, electric energy market modeling, and smart grid control are among the most topical and successful MATs. The described MAS illustrates how MATs enable adequate modeling of power-supplying devices with a non-trivial physical background. Such models of the lowest-level devices may be easily implanted into a model of a power system of any static or mobile power-consuming object, thus demonstrating an "additivity" of MATs, enabling their efficient application in multiple problem areas. Chapter 6 presents the useful application of ML to the development of one of the most interesting and practically valuable problems of operations research: approximate dynamic programming. It proposes an efficient ML algorithm for two-stage stochastic programs. The optimization framework makes it easy to introduce changes to the already obtained decisions as well as to capture the collective intelligence of the experienced decisions. Such features are hardly (or even not) available inside classic operations research approaches. Moreover, as computational results indicate, the proposed ML-based algorithm explicates rapid convergence.

Overall, this volume provides a comprehensive overview of the current state and perspectives of MATs–ML integration and application in a wide set of practical and theoretical problems.

Igor A. Sheremet
Full Member of the Russian Academy of Sciences,
Scientific Supervisor and Chair of the Science-Technological Board
of the National Computer Corporation (NCC),
Moscow, Russia

Section 1

Advanced Results in the Area of Integration of Multi-Agent Technologies and Machine Learning

Chapter 1

A State-of-the-Art Survey on Various Domains of Multi-Agent Systems and Machine Learning

Aida Huerta Barrientos and Alejandro Nila Luevano

Abstract

Multi-agent systems (MASs) are defined as a group of interacting entities or agents sharing a common environment that changes over time, with capabilities of perception and action, and the mechanisms for their coordination provide a modern perspective on systems that traditionally were regarded as centralized. The main characteristics of agents are learning and adaptation. In the last few years, MASs have received tremendous attention from scholars in different fields. However, there are still challenges faced by MASs and their integration with machine learning (ML) methods. The primary goal of the study is to provide a broad review of the current developments in the field of MASs combined with ML methods. First, we present features of MASs considering the ML perspective. Second, we provide a classification of applications of MASs combined with ML methods. Third, we present a density map of applications in E-learning, manufacturing, and commerce. We expect this study to serve as a comprehensive resource for researchers and practitioners in the area.

Keywords: machine learning, multi-agents, simulation, optimization, neural networks

1. Introduction

At the beginning of the 1990s, agent-based programming became an important part of simulation [1]. Although there is currently no formal definition of what an agent is, the term is used to describe self-contained programs that can control their own actions based on perceptions of their operating environment [2]. The goal of agent-based programming is to create programs that intelligently interact with their environment. The software used to program agents has its origins in the field of artificial intelligence, especially in the subfield of distributed artificial intelligence [3, 4], whose objective is the study of the properties of agents and the design of interaction networks between them.

In the same 1990s, as is suggested in Ref. [5], computational agents are typically characterized by:

- Autonomy: The agents had direct control of their actions and their internal state.

- Social skills: Agents interacted with other agents through a computational language.
- Reaction: Agents were able to perceive their environment and respond to it. The environment could be the physical world, a virtual world, or a simulated world that includes other agents.
- Proactivity: Because agents reacted to their environment, they themselves had to take goal-oriented initiative.

A typical agent-based model contains the following four elements [6]:

- Agents: Their attributes and environment.
- Relationships between agents and methods of interaction.
- A connectivity topology that defines how and with whom agents interact.
- Agent environments: Agents live and interact with their environment and with other agents.

In most of the agent-based models, agents are able to move within their environment through sensors through which they perceive their local neighbors. Communication between agents is usually done by sending messages. For this action, the agents must be able to listen to the messages that come from their environment and send messages to the environment.

1.1 Multiagent systems

According to Garro et al. [7], a system comprising a number of possibly interacting agents is called a *multiagent system* (*MAS*). In this direction, *MASs* are defined as a group of multiple autonomous interacting entities or agents sharing a common environment that changes over time, with capabilities of perception and action, and the mechanisms for their coordination provide a modern perspective on systems that traditionally were regarded as centralized. The main characteristics of agents are learning and adaptation. The main feature achieved when developing multi-agent systems is flexibility, since a multi-agent system can be added to, modified, and reconstructed, without the need for detailed rewriting of the application [7]. Cooperation is other feature that has been studied in *MASs*. In this direction, as suggested by Al-Jumaily and Al-Jaafreh [8], the *MASs* are divided into theory and application phases. On the one hand, taxonomy, cooperation structure, cooperation forming procedure, and others are related to the theory phase. On the other hand, mobile agent cooperation, information gathering, sensor information, and communication are related to the applications phase.

1.2 Machine learning

Machine learning (ML) is a subset of artificial intelligence (AI) that concerns the development of algorithms, which allows the machine to learn via inductive inference based on observation data that represent incomplete information about statistical phenomena [9]. To carry out the learning process an algorithm is used based on examples of the task we want to solve (data) and letting the computer find patterns and make inferences that optimize the decision-making according to a user-defined objective [10]. Based on the training strategy, ML can be divided into three classical categories with different learning approaches: supervised learning, unsupervised learning, and reinforcement learning [10]. The first one includes classification and regression tasks, in the second one the widely used task is clustering, and the third one consists of the process of training a model on a series of actions that lead to a particular outcome, where the system receives rewards for performing well and punishment for performing poorly directly from its environment [10].

In the last years, MASs integrated with ML have received tremendous attention from scholars in different fields such as Computer Science, Engineering, Mathematics, Material Science, Neuroscience, Energy, Physics and Astronomy, Social Sciences, Environmental Sciences, Business, Management, and Accounting. The overview of MASs integrated with ML in these fields will be presented in the following sections. MASs have been used in areas of e-learning, manufacturing, and commerce combining mathematical methods, optimization methods, Markov processes, learning algorithms, and artificial intelligence techniques. The reinforcement learning method jointly with MASs has been applied in e-learning, manufacturing (multi-agent reinforcement learning), and commerce (machine learning and learning automation) areas. While deep reinforcement learning and adaptive learning have been applied jointly with MASs in manufacturing and commerce areas. However, there are still challenges faced by MASs and their integration with ML. The primary goal of the study is to provide a broad review of the current developments in the field of MASs combined with ML methods.

This chapter is prepared as follows: the methodology followed to carry out the state-of-the-art survey on various domains of multi-agent systems and machine learning is described in Section 2. Features of MASs considering the ML perspective are presented in Section 3. A density map of applications in e-learning, manufacturing, and commerce is provided in Section 4. Concluding remarks are drawn in Section 5.

2. Methodology

We followed the methodology proposed by Machi and McEvoy [11]. **Figure 1** shows the steps for conducting the systematic literature review.

2.1 Step 1. Select a topic

The topic for this study was MASs combined with ML methods.

2.2 Step 2. Search the literature

The data for this study was extracted from the Scopus database (accessed on July 2022) based on the string *multi-agent* AND *systems* AND *machine* AND *learning*. In this case, 2869 relevant results were found. Then, the publication date of documents was limited from 2017 to 2022, the last five years. Also, the document type was limited to articles and book chapters. In this direction, the search was based on the following string:

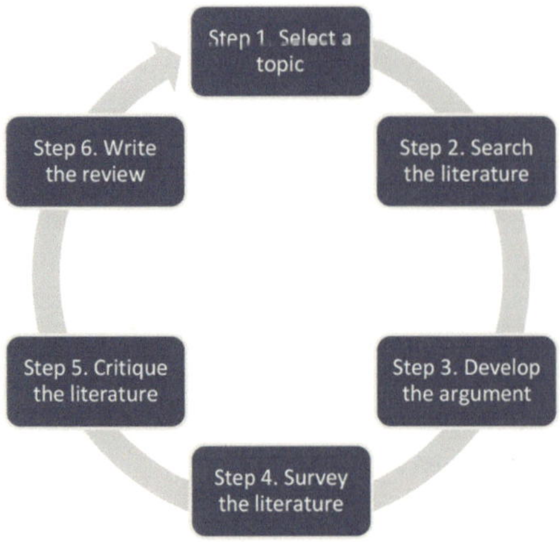

Figure 1.
The literature review model. Machi and McEvoy [11].

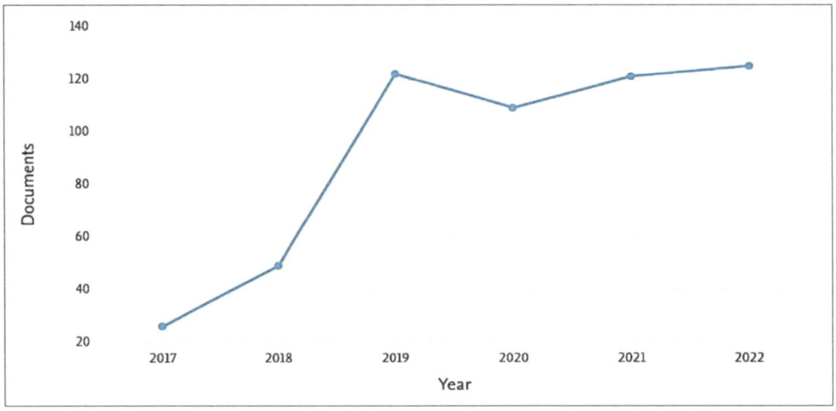

Figure 2.
Trends in the number of articles and chapter book by year that were published between 2017 and 2022.

TITLE-ABS-KEY (*multiagent* AND *systems* AND *machine* AND *learning*) AND (LIMIT-TO (PUBYEAR, *2022*) OR LIMIT-TO (PUBYEAR, *2021*) OR LIMIT-TO (PUBYEAR, *2020*) OR LIMIT-TO (PUBYEAR, *2019*) OR LIMIT-TO (PUBYEAR, *2018*) OR LIMIT-TO (PUBYEAR, *2017*)) AND (LIMIT-TO (DOCTYPE, "*ar*") OR LIMIT-TO (DOCTYPE, "*ch*")). In this case, 552 relevant articles and book chapters were found. All the information was exported in RIS format to VOSviewer software [12, 13] to generate the author collaboration and word co-occurrence networks. **Figure 2** shows the trends in the number of articles and book chapters published annually between 2017 and 2022. **Figure 3** depicts the fluctuating trend in the number of articles and chapter book published annually by the top-five sources, which reached a peak in 2019 by IEEE Access, and from then on it has gradually descended.

2.3 Step 3. Develop the argument

In this chapter, a broad review of the current developments in the field of MASs combined with ML methods is presented.

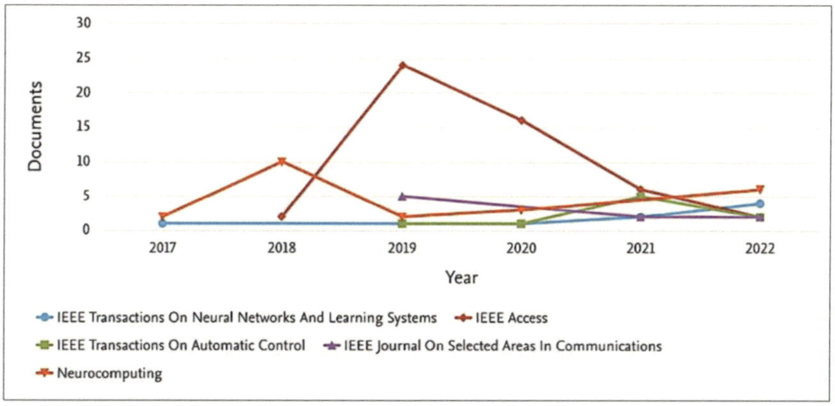

Figure 3.
Trends in the number of articles and chapter book published annually by the top-five sources.

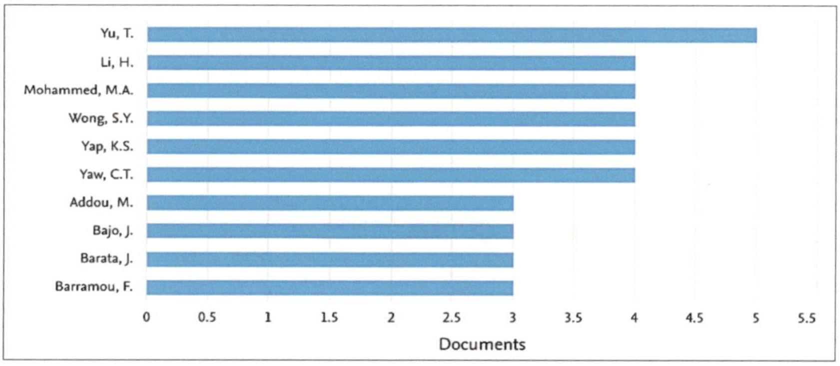

Figure 4.
Documents by the author published between 2017 and 2022 on multi-agent systems and machine learning applications.

2.4 Step 4. Survey the literature

Figure 4 summarizes the top ten authors in terms of contribution to the number of papers and chapter book published on multi-agent systems and machine learning applications. Even though Yu [14–18] as coauthor is the main contributor in the area. He accounts for five articles, followed by Li [19–22], Mohammed [23–26], Wong [27–30], Yap [27–30], and Yaw [27–30] with four articles each one. **Figure 5** presents the coauthor collaboration network from articles and chapter book published during the 2017–2022 period time, on multi-agent systems and machine learning applications, based on full counting. Each circle represents an author. The size of a circle reflects the number of publications of the corresponding author. The distance between two circles indicates the relatedness of the authors [13]. Colors represent clusters of authors with strong coauthorship links. A total of 1,718 different authors are in the collaboration network.

The top-ten fields in terms of applications on multi-agent systems and machine learning applications during the 2017–2022 period is showed in **Figure 6**. Even though Computer Science is the main field contributor, it only accounts for 37.4%, followed by Engineering (26.9%) and Mathematics (7.8%). The visualization

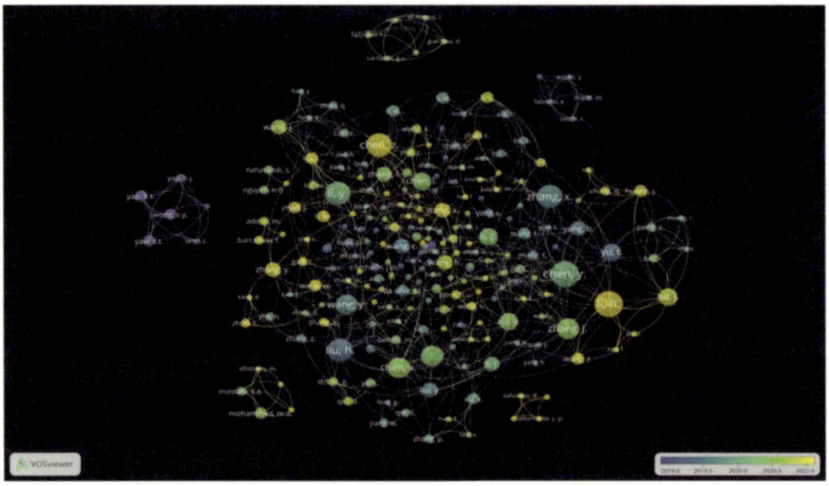

Figure 5.
Visualization by VOSviewer™ software of coauthor collaboration network from documents published between 2017 and 2022 on multi-agent systems and machine learning applications.

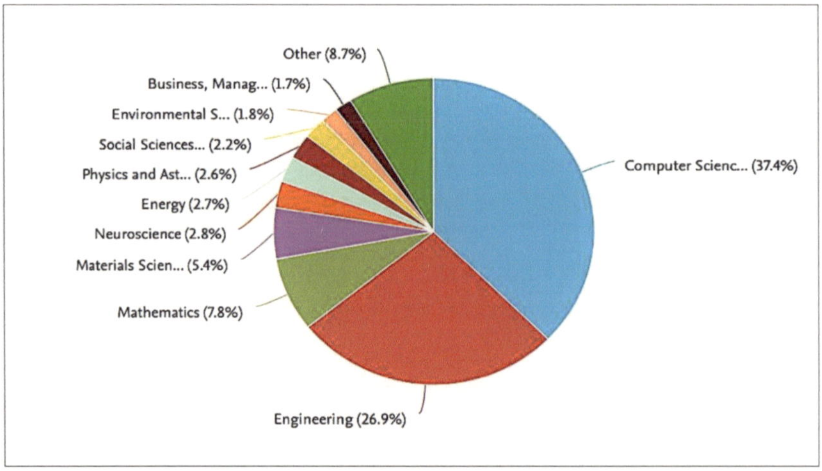

Figure 6.
Percentage of documents by subject fields that were published between 2017 and 2022.

in **Figure** 7, based on full counting, shows the co-occurrence network, distinct groups of keywords can be easily distinguished. Each circle represents a keyword. The size of a circle reflects the number of occurrences of the corresponding keyword. A total of 5,155 different words presented in the titles and abstracts of the 552 documents published between 2017 and 2022 were analyzed to establish the co-occurrence network, generating clusters associated with the research topic on multi-agent systems and machine learning applications. The colors used indicate the evolution of the time period of the different clusters. The blue cluster contained research topics published in 2019, while the green cluster contained research topics published in 2020, and the yellow cluster contained research topics published in 2021 and forward.

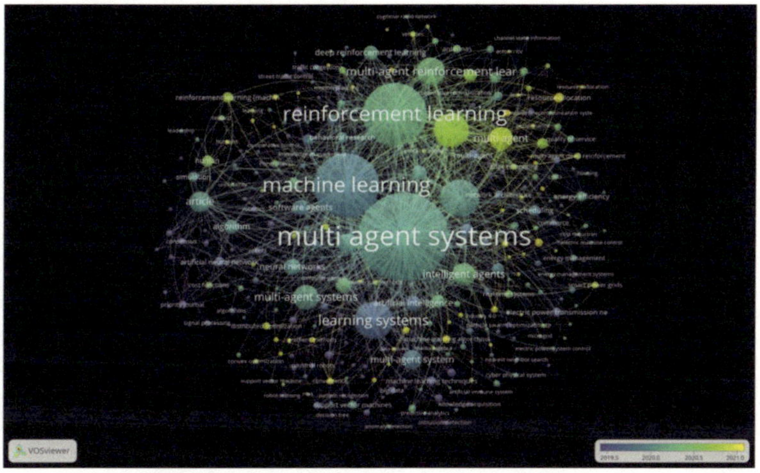

Figure 7.
Visualization by VOSviewer™ software of word co-occurrence network built using words present in titles and abstracts of documents published between 2017 and 2022 on multi-agent systems and machine learning applications.

2.5 Step 5. Critique the literature

The critique of the scientific literature of MASs integrated with ML in Computer Science, Engineering, Mathematics, Material Science, Neuroscience, Energy, Physics and Astronomy, and Social Sciences fields will be presented in the following sections, highlighting their potentiality and limitations.

3. Review of state-of-the-art

3.1 Computer science

Computer Science turned out as the top field contributor as indicated by the retrieved data, with 440 documents on multi-agent systems and machine learning applications during the 2017–2022 time period. **Figure 8** shows the terms with high co-occurrence frequencies in this field, distinguishing a color pallet from blue to yellow based on the publication year. In **Figure 8**, each term is represented by a circle, where the diameter of the circle and size of its label represents the frequency of the term, and its proximity to another term indicates the degree of relatedness of the two terms.

3.1.1 Potentiality

The network in **Figure 8** clearly indicates that multi-agent systems and machine learning have received tremendous attention from scholars. In 2019, the research was focused on topics such as embedded systems, robots, and forecasting, whereas in 2020, it was focused on topics such as intelligent agents, decision-making systems, optimization, IoT, stochastics systems scheduling, wireless sensor networks, compute games, and energy and efficiency of systems, which were broadly studied. More

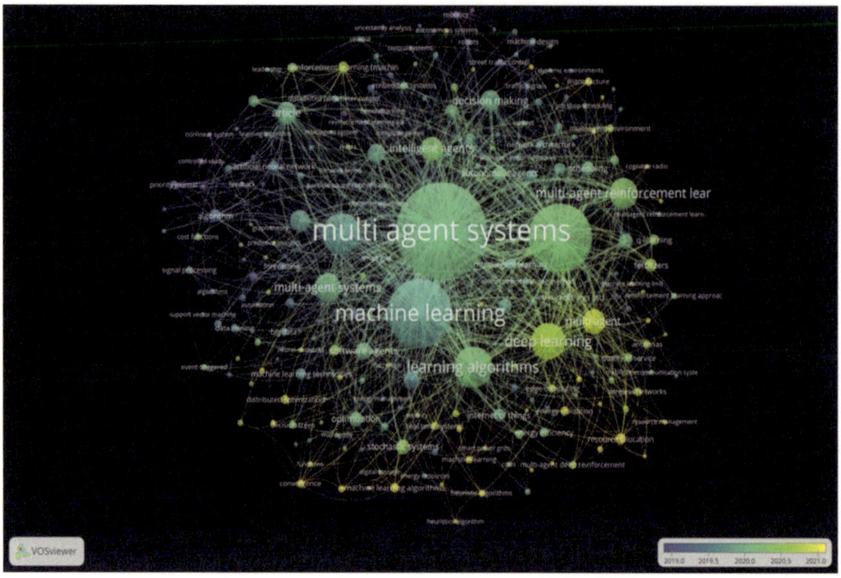

Figure 8.
Visualization by VOSviewer™ software of word co-occurrence network built using words present in titles and abstracts of documents published between 2017 and 2022 on multi-agent systems and machine learning applications in the Computer Science field.

recently, in 2021 resource allocation, heuristics algorithms, decision trees, real-time systems, distributed optimization, and predictive analysis jointly with multi-agent systems were applied to study street traffic, manufacture, and cognitive radio.

3.1.2 Limitations

ML algorithms and multi-agent systems showed potentially significant combination/integration in the Computer Sciences field. However, the studies in this field had several limitations. First, they had a small number of applications. Second, the ML algorithms are limited to deep reinforcement. Finally, it is unclear which software is used to integrate multi-agent systems and ML algorithms.

3.2 Engineering

Engineering turned out as the second contributor as indicated by the retrieved data, with 317 documents on MASs and ML applications between 2017 and 2022. **Figure 9** shows the terms with high co-occurrence frequencies in this field. In **Figure 9**, each term is represented by a circle, where the diameter of the circle and size of its label represent the frequency of the term, and its proximity to another term indicates the degree of relatedness of the two terms.

3.2.1 Potentiality

As is depicted in **Figure 9**, MASs researchers are very enthusiastic about learning algorithms to study vehicular networks, traffic control, 5G mobile communications networks, image communication systems, multi-robots systems, wireless sensor

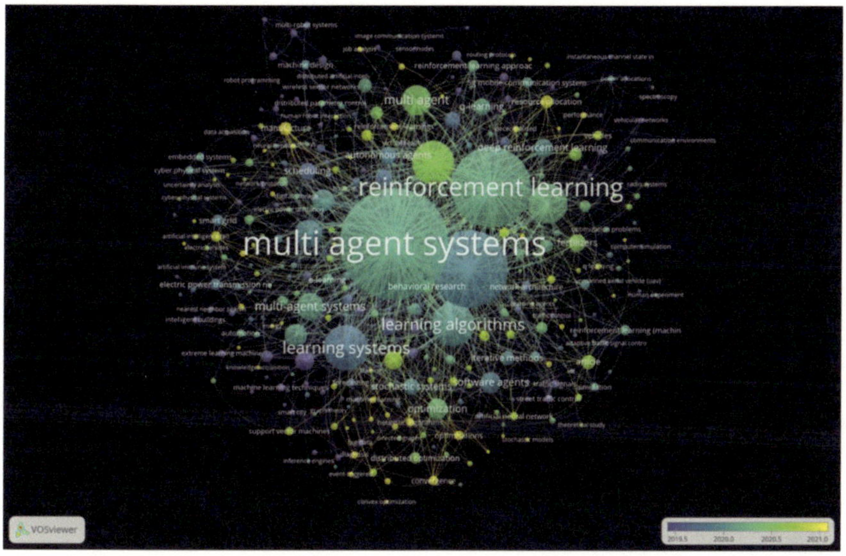

Figure 9.
Visualization by VOSviewer^{TM} software of word co-occurrence network built using words present in titles and abstracts of documents published between 2017 and 2022 on multi-agent systems and machine learning applications in the Engineering field.

networks, manufacture, digital storage, electric power utilization, cyber-physical systems, embedded systems, electric vehicles, electric power transmission, and e-learning, and have proposed reinforcement learning and deep reinforcement learning to be integrated to MASs.

3.2.2 Limitations

Although, MASs interacting with reinforcement learning and deep reinforcement learning algorithms showed potentially significant application in the Engineering field. However, there were several potential limitations, such as small minority of optimization techniques, traditional simulation approaches, and little information about the integration of optimization algorithms and simulation software.

3.3 Mathematics

Mathematics field turned out as the third contributor as indicated by the retrieved data, with 92 documents on MASs and ML applications between 2017 and 2022.

3.3.1 Potentiality

The number of publications increased noticeably in the past two years. The scientific landscape of main research areas of MASs and ML applications in the Mathematics field are bibliometrically explored by way of co-occurrence term map presented in **Figure 10**. In **Figure 10**, each term is represented by a circle, where the diameter of the circle and size of its label represent the frequency of the term, and its proximity to another term indicates the degree of relatedness of the two terms.

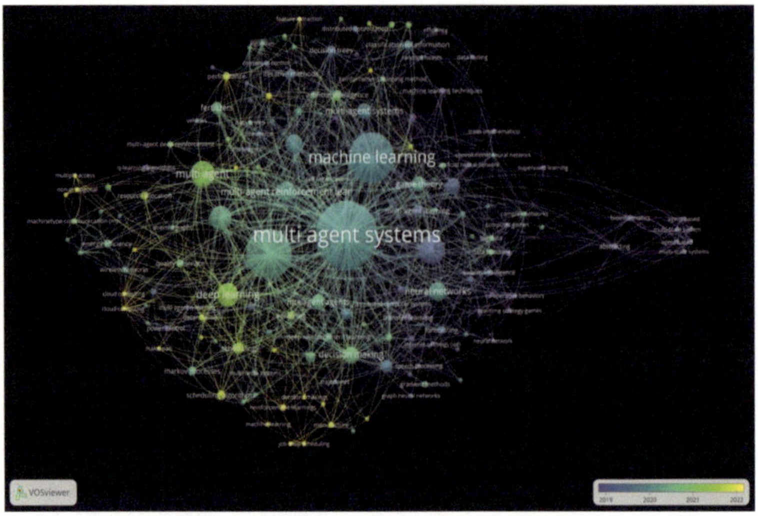

Figure 10.
Visualization by VOSviewer™ software of word co-occurrence network built using words present in titles and abstracts of documents published between 2017 and 2022 on multi-agent systems and machine learning applications in the Mathematics field.

The most used theoretical tools were deep learning, neural networks, intelligent agents, Markov process, scheduling algorithms, q-learning algorithms, decision trees, and game theory. While the most important application areas were decision making, job scheduling, power control, cloud computing, energy efficiency, fertilizers, e-commerce, speech processing, cooperative behaviors, quality of service, resource allocation, forecasting, and manufacturing.

3.3.2 Limitations

A considerable amount of literature has been published on multi-agent systems and machine learning applications in the Mathematics field. However, the reinforcement learning method has been recently used in specific areas such as manufacturing and job scheduling supporting decision making. Optimization algorithms based on warm intelligence were used in the past as well as supervised learning methods. Much of the recent literature in this field has limited applications that are centered on cloud computing.

3.4 Materials science

Materials Science field turned out as the fourth contributor as indicated by the retrieved data, with 63 documents on MASs and ML applications between 2017 and 2022. **Figure 11** shows the mapping and clustering of terms over time. A co-occurrence network was built using words present in titles and abstracts of documents published between 2017 and 2022 on multi-agent systems and machine learning applications in the Materials Science field. In **Figure 11**, each term is represented by a circle, where the diameter of the circle and size of its label represent the frequency of the term, and its proximity to another term indicates the degree of relatedness of the two terms. The network can be seen to contain four clusters of co-occurring terms.

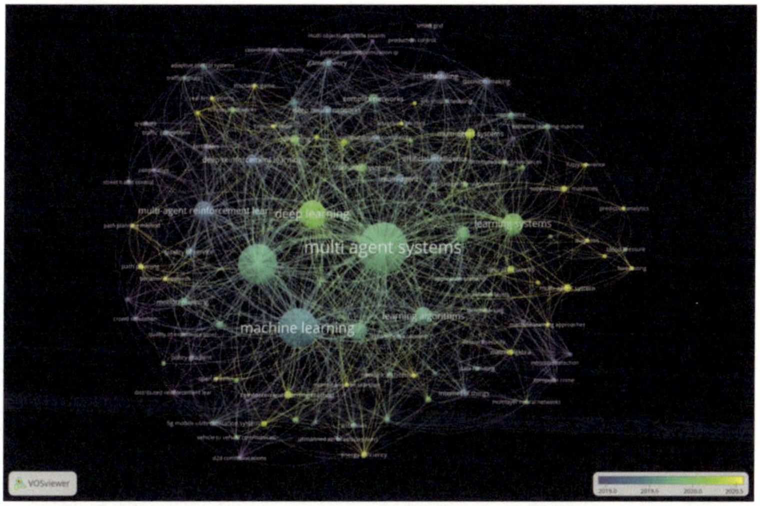

Figure 11.
Visualization by VOSviewer™ software of word co-occurrence network built using words present in titles and abstracts of documents published between 2017 and 2022 on multi-agent systems and machine learning applications in the Materials Science field.

The royal blue cluster is predominately associated with documents published in 2019, the turquoise cluster is associated with documents published in 2020, the yellow-green cluster is associated with documents, and the yellow cluster is associated with documents recently published in 2022.

3.4.1 Potentiality

For the period of 2017–2022, multi-agent reinforcement learning, deep learning, deep reinforcement learning, complex networks, multi-agent systems, stochastic systems, and game theory remained among the top major topics. It has been demonstrated that deep reinforcement learning methods have been combined with game theory and deep neural network techniques. Additionally, complex networks have considered computational complexity.

3.4.2 Limitations

Although there are many studies, the research in multi-agent systems and machine learning applications in the Materials Science field remains limited. First, the application areas have been centered on traffic studies, scheduling, Internet of Things, crowd simulation, and more recently in energy efficiency and wireless networks. Second, the learning algorithms are mainly based on deep learning. Additionally, optimization is based on the particle swarm algorithm, mainly.

3.5 Neuroscience

Neuroscience field turned out as the fifth contributor as indicated by the retrieved data, with 33 documents on MASs and ML applications between 2017 and 2022. **Figure 12** depicts the terms with high co-occurrence frequencies in this field.

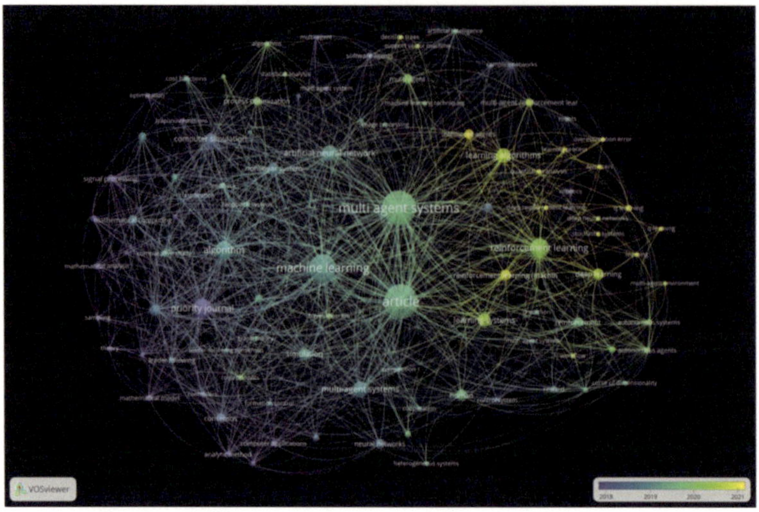

Figure 12.
Visualization by VOSviewer™ software of word co-occurrence network built using words present in titles and abstracts of documents published between 2017 and 2022 on multi-agent systems and machine learning applications in the Neuroscience field.

In **Figure 12**, each term is represented by a circle, where the diameter of the circle and size of its label represent the frequency of the term, and its proximity to another term indicates the degree of relatedness of the two terms.

3.5.1 Potentiality

The number of publications increased noticeably in the past two years on multi-agent systems and machine learning applications in the Neuroscience field. The terms on the left of the network shown in **Figure 12** represent the contributions developed in 2018. Here, mathematical models, simulation models, as well as analytic and optimization methods were broadly used. The group of terms at the center of the diagram characterizes the contributions developed between 2019 and 2020. Here, artificial neural networks, nonlinear systems, graph theory, and simulation were the theoretical tool mainly used. The right of the diagram contains terms related to contributions in the last year. Here, the learning and reinforcement algorithms have been broadly applied.

3.5.2 Limitations

ML algorithms and multi-agent systems showed potentially significant combination/integration in the Neurosciences field. However, the studies in this field had several limitations. First, the applications are centered on autonomous systems that learn using deep learning and reinforcement learning algorithms. Second, the reward concept is minimum exploited in reinforcement learning. The software to implement multi-agent systems has little attention. Finally, computer simulation was relevant in studies developed in 2018 but is not included in the studies recently developed.

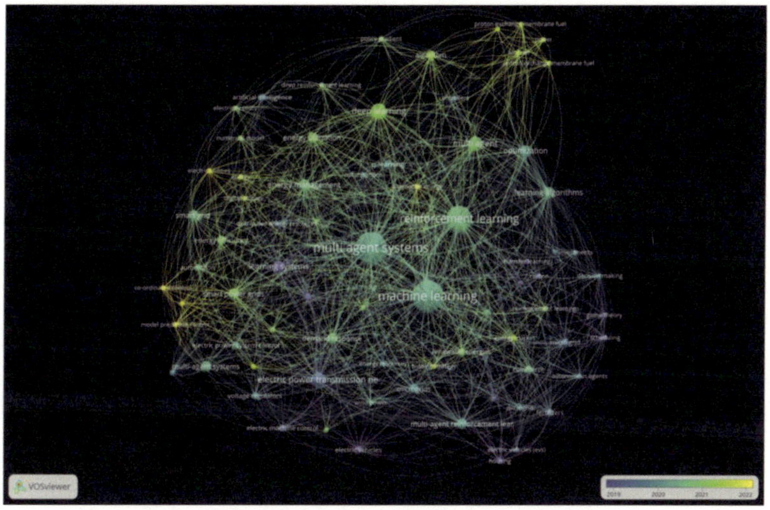

Figure 13.
Visualization by VOSviewer™ software of word co-occurrence network built using words present in titles and abstracts of documents published between 2017 and 2022 on multi-agent systems and machine learning applications in the energy field.

3.6 Energy

Energy field turned out as the sixth contributor as indicated by the retrieved data, with 32 documents on MASs and ML applications between 2017 and 2022. **Figure 13** shows the terms with high co-occurrence frequencies in this field. In **Figure 13**, each term is represented by a circle, where the diameter of the circle and size of its label represent the frequency of the term, and its proximity to another term indicates the degree of relatedness of the two terms.

3.6.1 Potentiality

A number of specific research topics show significant active growth and may be considered to be emerging topics on multi-agent systems and machine learning applications in the Energy field. Reinforcement learning is the dominant algorithm used followed by deep learning and deep reinforcement learning. Optimization is based on learning algorithms. The main applications in this field are electric machine control, electric vehicles, housing, fertilizers, smart grid, electric power transmission networks, microgrids, energy management, and energy utilization.

3.6.2 Limitations

There are two major limitations in the applications of multi-agent systems and machine learning in the Energy field. First, the more recent applications are centered on model predictive control. Second, the electric cost is evaluated in a few studies.

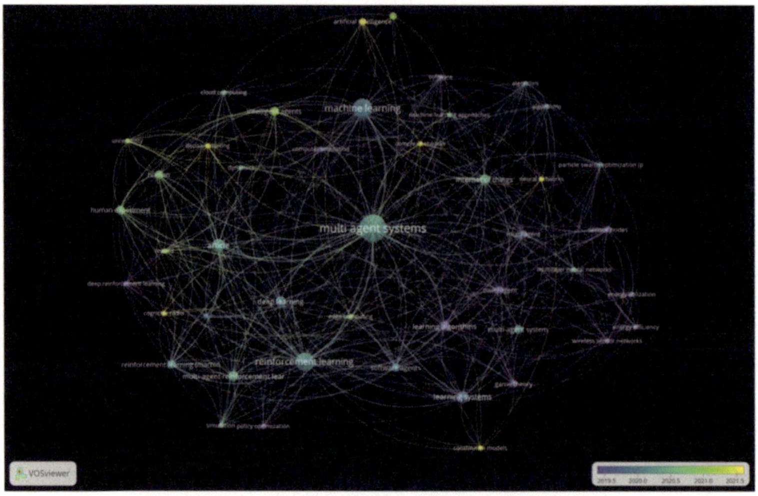

Figure 14.
Visualization by VOSviewer™ software of word co-occurrence network built using words present in titles and abstracts of documents published between 2017 and 2022 on multi-agent systems and machine learning applications in the Physics and Astronomy field.

3.7 Physics and astronomy

Physics and Astronomy field turned out as the seventh contributor as indicated by the retrieved data, with 31 documents on MASs and ML applications between 2017 and 2022. **Figure 14** illustrates the network of terms with high co-occurrence frequencies in this field. In **Figure 14**, each term is represented by a circle, where the diameter of the circle and size of its label represent the frequency of the term, and the distance between any two possible terms reflects the relatedness of the terms as closely as possible. In general, the stronger the relationship between two terms, the smaller the distance between the terms on the map. The network can be seen to contain four clusters of co-occurring terms.

3.7.1 Potentiality

After careful analysis, in 2019, three topics can be distinguished (the royal blue biggest circles in **Figure 14**). The first is learning algorithms, the second topic is sensor nodes, and the third topic is game theory.

In 2020, six topics (the turquoise biggest circles in **Figure 14**) were the most often discussed of the last three years. The observation that reinforcement learning is one of the most occurring terms in the Physics and Astronomy field does not come as a surprise Internet of Things.

In 2021, six topics (the yellow biggest circles in **Figure 14**) were the most predominant of the last three years. The prominence of neural networks may be explained by the fact that is linked directly to machine learning approaches, constituent models, and multilayer neural networks.

3.7.2 Limitations

Although, MASs interacting with reinforcement learning and deep reinforcement learning algorithms showed potentially significant application in Physics and Astronomy field. However, there were several potential limitations, such as small minority of simulation and optimization models, learning algorithms, and software.

4. Applications of MASs combined with ML

Density map assists in the understanding of active growth areas, research trends, emerging topics, and hot topics in MASs combined with ML. **Figures 15–17** illustrate the density map of applications in E-learning, manufacturing, and commerce, respectively.

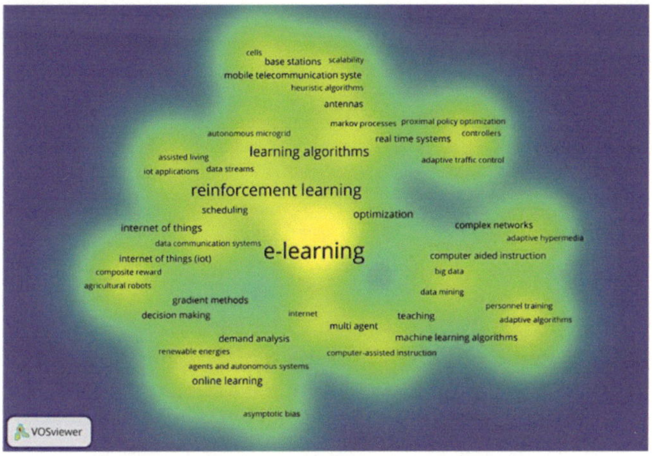

Figure 15.
Density map by VOSviewer^{TM} software of word co-occurrence network built using words present in titles and abstracts of documents published between 2017 and 2022 on multi-agent systems and machine learning applications in E-learning.

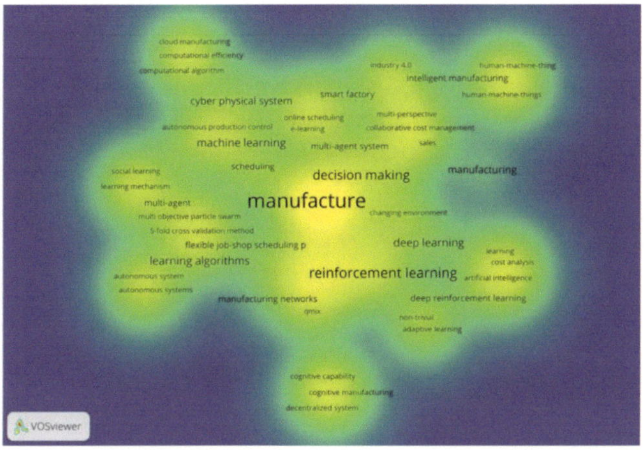

Figure 16.
Density map by VOSviewer^{TM} software of word co-occurrence network built using words present in titles and abstracts of documents published between 2017 and 2022 on multi-agent systems and machine learning applications in manufacturing.

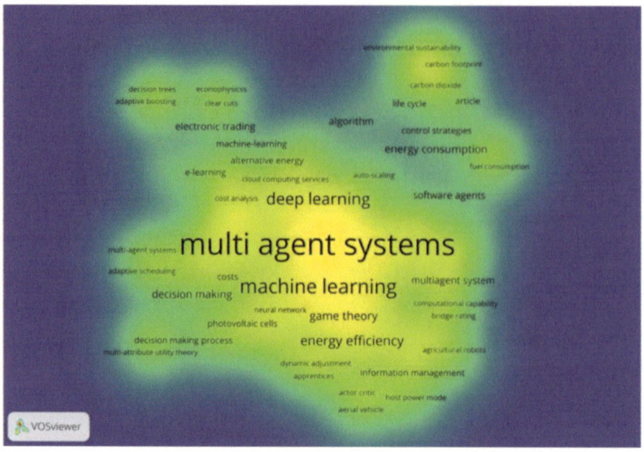

Figure 17.
Density map by VOSviewer™ software of word co-occurrence network built using words present in titles and abstracts of documents published between 2017 and 2022 on multi-agent systems and machine learning applications in commerce.

5. Conclusions

The primary goal of the study was to provide a broad review of the current developments in the field of MASs combined with ML methods. The trend on MASs and ML is the use of reinforcement learning algorithms integrated with optimization and simulation models. Artificial neural networks, nonlinear systems, and graph theory are the theoretical tool mostly used. We want to emphasize that it is unclear which software is used to integrate multi-agent systems and ML algorithms. In most studies, optimization algorithms were based on warm intelligence.

Acknowledgements

The authors appreciate the partial support by UNAM- PAPIME PE 104022. The authors also gratefully acknowledge all the time and support received from IntechOpen.

Conflict of interest

The author declares no conflict of interest.

Author details

Aida Huerta Barrientos* and Alejandro Nila Luevano
Faculty of Engineering, National Autonomous University of Mexico, Mexico City, Mexico

*Address all correspondence to: aida.huerta@comunidad.unam.mx

References

[1] Maes P. Agents that reduce work and information overload. Communications of the ACM. 1994;**37**:31-40

[2] Huhns M, Singh MP. Readings in Agents. 1st ed. San Mateo, CA: Morgan Kaufmann; 1998

[3] Bond AH, Gasser L. Readings in Distributed Artificial Intelligence. Los Altos, CA, USA: Morgan Kaufmann; 1998

[4] Chaib-draa B, Moulin B, Mandiau R, Millot P. Trends in distributed artificial intelligence. Artificial Intelligence Review. 1992;**6**:35-66

[5] Wooldridge M, Jennings NR. Intelligent agents: Theory and practice. The Knowledge Engineering Review. 1995;**10**:115-152

[6] Macal C, North M. Introductory tutorial: Agent-based modeling and simulation. In: Proceedings of the 2011 Winter Simulation Conference. Phoenix, Arizona, USA: IEEE; 2011. pp. 1468-1468

[7] Garro A, Mühlhäuser M, Tundis A, Baldoni M, Baroglio C, Bergenti F, et al. Intelligent agents: Multi-agent systems. Encyclopedia of Bioinformatics and Computational Biology. 2019;**1**:315-320

[8] Al-Jumaily A, Al-Jaafreh M. Multi-agent system concepts, Theory and Applications phases. In: Buchli J, editor. Mobile Robots, Moving Intelligence. London: InTech; 2006. pp. 369-392

[9] Cen L, Dong M, Li Zhu H, Yu L, Chan P. Machine learning methods in the application speech emotion recognition. In: Zhang Y, editor. Application of Machine Learning. London: InTech; 2010

[10] Haidine A, Zahra Salmam F, Aqqal A, Dahbi A. Artificial intelligence and machine learning in 5G and beyond: A survey and perspectives. In: Haidine A, editor. Moving Broadband Mobile Communications Forward. Intelligent Technologies for 5G. London: InTech; 2021

[11] Machi LA, Mcevoy BT. The Literature Review. California: Corwin Press; 2009

[12] Van Eck NJ, Waltman L. Software survey: VOSviewer, a computer program for bibliometric mapping. Scientometrics. 2010, 2010;**84**:523-538

[13] Van Eck NJ, Waltman L. Visualizing bibliometric networks. In: Ding Y, Rousseau R, Wolfram D, editors. Measuring Scholarly Impact: Methods and Practice. Belin: Springer; 2014. pp. 285-320

[14] Li J, Yang B, Yu T. Distributed deep reinforcement learning-based coordination performance optimization method for proton exchange membrane fuel cell system. Sustainable Energy Technologies and Assessments. 2022;**50**:101284

[15] Li J, Yu T. Large-scale multi-agent deep reinforcement learning-based coordination strategy for energy optimization and control of proton exchange membrane fuel cell. Sustainable Energy Technologies and Assessments. 2021;**48**:101752

[16] Chen Y, Zhang X, Guo L, Yu T. Optimal carbon-energy combined flow in power system based on multi-agent transfer reinforcement learning. Gaodianya Jishu/High Voltage Engineering. 2019;**45**:3

[17] Cheng L, Yu T. Exploration and exploitation of new knowledge

emergence to improve the collective intelligent decision-making level of web-of-cells with cyber-physical-social systems based on complex network modeling. IEEE Access. 2018;**6**

[18] Zhang X, Chen Y, Yu T, Yang B, Qu K, Mao S. Equilibrium-inspired multiagent optimizer with extreme transfer learning for decentralized optimal carbon-energy combined-flow of large-scale power systems. Applied Energy. 2018;**189**:157-176

[19] Liu C, Shi Y, Li H, Du W. Accelerated dual averaging methods for decentralized constrained optimization. IEEE Transactions on Automatic Control. 2022. DOI: 10.1109/TAC.2022.3173062

[20] Sun B, Hu J, Xia D, Li H. A distributed stochastic optimization algorithm with gradient-tracking and distributed heavy-ball acceleration. Frontiers of Information Technology and Electronic Engineering. 2021;**22**:11

[21] Li H, Hu J, Ran L, Wang Z, Lü Q, Du Z, et al. Decentralized dual proximal gradient algorithms for non-smooth constrained composite optimization problems. IEEE Transactions on Parallel and Distributed Systems. 2021;**32**:103

[22] Hu J, Ran L, Du Z, Li H. Decentralized stochastic optimization algorithms using uncoordinated step-sizes over unbalanced directed networks. Signal Processing. 2021;**32**:10

[23] Mohammed MA, Elhoseny M, Abdulkareem KH, Mostafa SA, Maashi MS. A multi-agent feature selection and hybrid classification model for parkinson's disease diagnosis. ACM Transactions on Multimedia Computing, Communications and Applications. 2021;**17**:2

[24] Mohammed MA, Ibrahim DA, Salman AO. Adaptive intelligent learning approach based on visual anti-spam email model for multi-natural language. Journal of Intelligent Systems. 2021;**30**:1

[25] Elhoseny M, Mohammed MA, Mostafa SA, Abdulkareem KH, Maashi MS, Garcia-Zapirain B, et al. A new multi-agent feature wrapper machine learning approach for heart disease diagnosis. Computers, Materials and Continua. 2021;**67**:1

[26] Mostafa SA, Mustapha A, Mohammed MA, Hamed RI, Arunkumar N, Abd Ghani MK, et al. Examining multiple feature evaluation and classification methods for improving the diagnosis of Parkinson's disease. Cognitive Systems Research. 2019;**54**:90-99

[27] Yaw CT, Yap KS, Wong SY, Yap HJ, Paw JKS. Enhancement of neural network based multi agent system for classification and regression in energy system. IEEE Access. 2020;**8**:163026-163043

[28] Yaw CT, Wong SY, Yap KS, Tan CH. Extreme learning machine neural networks for multi-agent system in power generation. International Journal of Engineering and Technology (UAE). 2018;**7**:4

[29] Yaw CT, Wong SY, Yap KS, Yap HJ, Amirulddin UAU, Tan SC. An ELM based multi-agent system and its applications to power generation. Intelligent Decision Technologies. 2018;**12**:2

[30] Yaw CT, Wong SY, Yap KS, Yap HJ, Amirulddin UAU, Tan SC. An ELM based multi-agent system and its applications to power generation. Intelligent Decision Technologies. 2017;**11**:3

Chapter 2

Deep Multiagent Reinforcement Learning Methods Addressing the Scalability Challenge

Theocharis Kravaris and George A. Vouros

Abstract

Motivated to solve complex demand-capacity imbalance problems in air traffic management at the pre-tactical stage of operations, with thousands of agents (flights) daily, even in a restricted airspace, in this paper, we review deep multiagent reinforcement learning methods under the prism of their ability to scale toward solving problems with large populations of heterogeneous agents, where agents have to unavoidably decide on their joint policy, without communication constraints.

Keywords: deep reinforcement learning, multiagent systems, scalability

1. Introduction

Scalability in training large numbers of deep reinforcement learning agents, which must decide on actions jointly, is a major issue that becomes apparent in many real-life problems. This issue is related to numerous aspects of deep multiagent reinforcement learning (DMARL), such as assignment of credits to the learners for their choices, assumptions regarding homogeneity or interchangeability of the agents, society structure due to interaction of agents' decisions, agents' communication requirements, abilities, and constraints.

In this chapter, we provide a review of deep multiagent reinforcement learning (DMARL) methods, examining their ability to scale up to large agent populations. "Large" here could mean anything from hundreds up to several thousands of agents.

We are motivated to solve complex demand-capacity balancing problems in air traffic management (congestion problems regarding air sectors), where we may have 6000 flights daily, even in a restricted airspace (e.g., the Spanish airspace). Specifically, in the air-traffic management (ATM) domain, *demand and capacity balancing* problems (DCBs) are a kind of congestion problems that arise naturally whenever demand of airspace use exceeds capacity, resulting to *"hotspots."* Hotspots are resolved via capacity management or flow management solutions, including regulations that generate delays and reroutings to flights, causing unforeseen effects for the entire system, and increasing uncertainty regarding the scheduling of (ground and airspace) operations. For instance, flight delays cause the introduction/increase of time buffers in operations' schedules and may accumulate demand for resources within specific periods. These are translated into costs and negative effects on airlines' reliability, customers' satisfaction, and environmental footprint. Representing flights by self-interested, heterogeneous agents (each with its own possible regulations, preferences, and constraints), the automation of problems' resolution requires agents to jointly decide on their policies regarding their own regulations [1]. Driven by the resolution of problems at the pre-tactical phase of operations (i.e., from some hours to few days before the actual flights), there are no communication or observation constraints for agents: In any case, agents need to coordinate with any agent with whom they participate to any specific congestion problem, given also that these problems may emerge dynamically in space and time.

In such large-scale, complex multiagent settings, we need to consider quality of solutions (as regulations incur additional costs to operations) and DMARL methods' training scalability. Factors that affect training scalability are as follows:

- The *training paradigm* adopted: Agents may train independent, centralized, or shared models.
- The *types of models* learnt following any paradigm: A policy model, a value model, or both types of models may be fit, following any of the training paradigms.
- *Assumptions regarding agents homogeneity*: Agents may be considered to be heterogeneous, homogeneous (i.e., following the same policy, which may be however differentiated due to contextual features), or even interchangeable.
- *Effectiveness of communication*: Communication between agents may be explicit (i.e., passing information by any means) or implicit (i.e., via the environment, or via sharing models' parameters), and orthogonal, either be performed in a global (e.g., broadcasting messages to all agents) or local scale (i.e., in a "neighborhood"). Here, optimality of communication is something that agents could learn via elaborated (e.g., attention) mechanisms.

In addition to the above factors, decomposing rewards among agents is a relevant issue, affecting both scalability and the quality of joint policies. This issue of credit assignment concerns designating high reward to the agents with a desirable behavior, thus avoiding agents enjoying the rewards without contributing in achieving the intended joint goal.

2. Deep MARL

Tabular function representations in reinforcement learning (RL) have many successes [2] in relatively low-dimensional problems, but it has two major drawbacks: (a) The designer of the application had to hand-craft the state representations, and (b) methods store each state or state-action value (V-value or Q-value, respectively) independently, resulting in slow learning in large state-action spaces and poor generalization abilities.

To resolve these problems, the deep Q-network [3] method successfully combined RL with neural networks (NNs). It is well established that the combination of RL with nonlinear function approximators such as NNs can be unstable and result in divergence [4]. This instability has several causes: the correlations present in the sequence of observations, the fact that small updates to Q values may significantly change the policy—and therefore change the data distribution of the samples produced by the rollouts, and the correlations between the action values and the target values. Deep Q-Network (DQN) [3] uses an NN to approximate *Q*-values, modeling the agent's policy.

Two vital elements of DQN that address these issues are the target network and the experience replay memory. The target network mitigates the effect of constantly moving targets, by incorporating a second network from which the targets are sampled. This network is periodically updated with the weights of the online network. The addition of a uniform experience replay memory decorrelates the samples collected during rollouts, by randomizing over the data, thereby smoothing over changes in the data distribution. During learning, the method applies Q-learning updates on samples (or minibatches) of experience drawn uniformly at random from the pool of samples stored.

Since the original DQN publication, many extensions have emerged [5–9], with double DQN [10] and prioritized experience replay [11] being the most well known. Double Q learning was originally introduced by H. Hasselt [12] and aims to address the problem of overestimating action values, a phenomenon inherent to the Q-learning method. This addition to the original method is considered to be standard practice and is particularly useful in the multiagent domain, where nonstationarity is a common phenomenon. Originally, the idea behind this method is to utilize two independent tabular representations of the Q function during training, where each Q function is updated with targets produced from the other Q function. Double DQN transfers this approach in the Deep RL setting, by exploiting the second approximator using a target network.

In the initial DQN approach, experience transitions are uniformly sampled from a replay memory. This approach ignores the significance of samples, replaying them at the same frequency that they were originally observed in the environment. Prioritized experience replay [11] enforces a priority over the samples, aiming to replay important transitions more frequently, and subsequently improve learning efficiency. In particular, the method proposes to assign higher priority to transitions with high expected learning progress, as measured by the magnitude of their temporal-difference (TD) error.

This new paradigm of combining Q-learning with NNs swiftly crossed over to the multiagent (MAS) field. Tampuu et al. [13] studied cooperation and competition between two DQN agents. The publication investigates the interaction between two agents in the well-known video game Pong, by utilizing two independent DQN architectures, one for each agent. By solely adjusting the rewarding scheme, competitive and collaborative behaviors emerge.

Deep deterministic policy gradient (DDPG) [14] combines DQN with deterministic policy gradient (DPG) [15] to address continuous actions. As a variant of actor-critic algorithms it incorporates two separate NNs: the actor, which models the policy, and the critic, which provides feedback on the desirability of an action a in a state s. This desirability can be expressed by means of learnt V-values, Q-values, or advantages. MADDPG [16] extends DDPG in MAS settings: Each agent is given a dedicated policy network. This approach renders the method not scalable: It is not viable to maintain and train hundreds, if not thousands, of policy networks. In addition, detrimental to the scalability is the fact that, during training, the method utilizes a centralized critic, which takes as input the observations and actions of all agents. The input size of the critic can explode, depending on the size of the agent population and the state dimensions.

Multiagent deep deterministic policy gradient (MADDPG) employs a technique that has many variants and is vital in understanding deep MARL. This technique is called centralized training and decentralized execution (CTDE). A naive way to provide agents with a joint policy is to train in parallel independent policies, one for each agent, following the independent learners paradigm. This approach can produce results in low-dimensional problems, but in real-world scenarios is inefficient and unstable. In order to alleviate these problems produced by the nonstationarity of an MAS environment, many algorithms, one of which is MADDPG, employ some form of centralized training, thus exploiting information that is available during training but often unavailable during execution. In particular, MADDPG trains a centralized critic, which takes as input the global state and agents' joint action. During the execution phase, this information is inaccessible by agents, and agents act in a decentralized manner. Another well-known application of CTDE is QMIX [17]. During training, the learning algorithm has access to all local action observation histories and global state, but each agent learns a policy that conditions actions on local agent observations.

Parameter sharing, first introduced by Gupta et al. [18], is the extreme case of CTDE: It learns a single policy shared by many agents. The main idea is that this single policy should be able to adequately describe the behavior of different agents with the same goals. Immediately apparent is the important assumption that the agents are homogeneous, meaning that they decide on the same state-action space. This assumption is relaxed by the same publication [18], which "tagged" observations with agent identifiers, allowing differentiating agents according to their goals, and responding accordingly. The resulting policy can be more robust, given the fact that it has been trained with samples that potentially belong to different parts of the state space, explored by different agents. We consider the parameter sharing approach to be the cornerstone of any scalable algorithm.

Castañeda proposes two multiagent variants of DQN [19]. Repeated update Q-learning extends DQN by updating each action inversely proportional to the probability of selecting it, practically changing the learning rate based on the action probability. It is designed to mitigate the inherit overestimation of state values by Q-learning, much like double Q-learning [10]. Loosely coupled Q-learning defines a diffusion function and utilizes eligibility traces in order to associate negative rewards with states. The combination of the diffusion function with eligibility traces is used to define a function, which indicates the necessity for cooperation, expressing the probability of an agent carrying on an action independently. It combines this with Dec-MDP and two Q-value functions, one for independent acting (when appropriate) and one for coordinating with others.

Credit assignment concerns the difficulty to give credit and provide higher reward to the agents with a desirable behavior, thus accelerating learning in MAS settings. A common phenomenon, called "lazy agent," occurs when an agent does not participate actively in a cooperative solution, enjoying the rewards resulting from the cooperation of others. Various approaches have been proposed to deal with this problem, with the difference reward [20] and reward shaping [21, 22] being the most well known.

Difference reward [20] introduces a method to visualize the desirable properties of a reward function, thus facilitating the creation of new reward structures based on the specific needs of the domain. Specifically, the authors in [20] focus on two aspects of the reward function. The first is called *factordness* and expresses how well the reward promotes coordination among agents in different parts of a domain's state-space. The second is called *learnability* and expresses how easy it is for an agent to learn to maximize its reward, by measuring the reward's sensitivity to the agent's actions. The main idea behind reward shaping is to utilize some prior knowledge while engineering the reward function, in order to reduce the training time, by reducing the number of suboptimal actions taken [22]. In the most general form, reward shaping can be as simple as $R' = R + F$, where R is the original reward and F a positive scalar reward, designed to encourage the agent to move toward the goal. Here, most of the research focuses on the principles, which would lead to an effective reward design, as well as how the optimal policy changes as an effect of reward design.

Value decomposition network (VDN) [23] is the first DMARL method that addresses the credit assignment problem. VDN represents joint action value as a summation of local (individual agents') action values conditioned on agents' local observations. The fact that the agents' local value function depends only on local observations facilitates agents' understanding of the received rewards. This method is not scalable, because it utilizes distinct NNs for each agent's policy. QMIX [17] provides a more general case of VDN using a mixing NN that combines all independent Q-values into Q_{tot}, thus approximating a broader class of functions for joint action values. Specifically, moving from the VDN's assumption of additivity, QMIX's mixing network can approximate monotonic relations between individual and the global Q-values. Both works are based on the individual-global-max (IGM) principle, according to which, the joint greedy action should be equivalent to the set of individual greedy actions of agents.

VDN and QMIX address a subset of factorizable MARL tasks due to their structural constraint in factorization of additivity and monotonicity, respectively. Later works have improved the representation ability of the mixing network. QTRAN [24] achieves more general factorization by transforming the original joint action-value function into a new, easily factorizable one with the same optimal actions as the original. NDQ [25] combines value function factorization learning with communication learning, by introducing an information-theoretic regularizer, aiming to reduce interagent communication while maintaining performance.

ROMA [26] combines MARL, mixing networks in particular, with the role concept [27–33]. A role is a comprehensive pattern of behavior, often specialized in some tasks. Agents with similar roles will show similar behaviors and thus can share their experiences to improve performance. The main drawback of this approach is the demand of exploitation of prior domain knowledge in order to decompose tasks and predefine the responsibilities of each role, which necessitates adding to role-based MAS dynamic and adaptive abilities to perform effectively in dynamic and unpredictable settings. However, the specification of roles may not be appropriate for any domain. ROMA introduces two regularizers to enable roles to be dynamically identifiable, in order to exploit the benefits of both, role-based and learning paradigms. Following the QMIX [17] framework, it utilizes independent policy networks and a centralized mixing network.

QPLEX [34] focuses on ensuring that the IGM principle (as specified above) stands, while reformalizing it as an advantaged-based IGM. QPLEX replaces the original mixing network of QMIX [17] with a duplex dueling network architecture [5], which induces the joint and local (duplex) advantage functions, to factorize the joint action-value function into individual action-value functions. This duplex dueling structure encodes the IGM principle into the neural network architecture. The architecture uses an individual action-value function for each agent in combination with the centralized duplex dueling component. The method manages to achieve higher win rates in numerous Starcraft II scenarios [35] against multiple baselines such as VDN, QMIX, and QTRAN [17, 23, 24] in experiments with up to 27 agents.

Another well-known approach is the counterfactual multiagent policy gradients method (COMA) [36]: It decomposes the global reward to the agents, utilizing a counterfactual baseline inspired by difference rewards [20]. Similarly to MADDPG, COMA utilizes a centralized critic and each agent has a distinct policy network, so the drawbacks regarding scalability are common to MADDPG.

Concluding the above, the line of research on value decomposition has different focal points rather than scalability, thus does not produce methods ideal to scale up to thousands of agents. In addition to that, as far as credit assignment problem is concerned, although a vital aspect of MARL, we do not consider that the resolution of this problem is a prerequisite toward methods' scalability. However, it is a related issue in settings where rewards are not inherently decomposed to individual agents.

3. Scalable deep MARL

Differentiable interagent learning (DIAL) [37] is the first proposal for learnable communication utilizing DQNs. Agents generate real-valued messages, which are broadcasted to the other agents. During centralized training, these messages are practically gradients, allowing end-to-end training across agents. During decentralized execution, messages are discretized and mapped to a predefined set of communication actions. There are two major reasons that this approach can not scale effectively: With no parameter sharing in place, each agent uses its own, independent policy NN. Also, every agent communicates with everyone else throughout training. This, beyond the communication cost, could result in agents receiving misleading or noisy information, in the form of gradients.

CommNet [38] aims to learn a communication protocol alongside the policies of cooperating agents. CommNet consists of a centralized feed-forward NN, with the observations of all agents as input and their actions as output. Each layer represents a communication step between the agents and consists of one decision module per agent. While the first layer uses an encoder function, every hidden layer module takes the internal state h as well as the broadcasted communication vector c as input and outputs the next internal state. These internal states are averaged and broadcasted as the next communication vector. CommNet has three major limitations regarding scalability. First, all agents use the same centralized network, which has to be big enough to accommodate this design approach, due to the size of the input in the form of the global state. In addition, this renders the method unsuitable for heterogenous agents. Second, CommNet calculates the mean of messages between layers, therefore assuming that all agents are of equal importance, which could be unsuitable in some settings. Finally, the most important disadvantage is the fact that all agents have to communicate with everyone. This could result in significant communication overhead with noise in the communication channel: e.g., an agent could receive a communication vector consisting of the average observations of irrelevant agents. The last drawback could be mitigated by the local connectivity model extension proposed in the original CommNet publication [38]. According to this extension, agents can only receive messages from a dynamic group of agents, which are within a certain range. However, no experiments are provided for this extension.

Toward resolving the mandatory global communication problem, an extension of CommNet, called vertex attention interaction network (VAIN) [39], employs an attention mechanism. This attention mechanism improves the performance of the original method by modeling the locality of interactions and thus determining which agents will share information. The authors claim that this method can make CommNet suitable for as many as 1000 agents. In the experiments provided, VAIN achieved better score than CommNet, although for a small number of agents.

Another method with scalability potential is BiCNet, [40]. BiCNet is an extension of the actor-critic formulation where agents use a bidirectional recurrent NN (RNN) [41]. The recurrent network serves as a bidirectional communication channel but also as a local memory saver: Each agent is able to maintain its own internal states, as well as to share information with its neighbors. Similarly to CommNet, the communication channel serves to broadcast agents' local observations. Contrary to CommNet agents, BiCNet agents utilize additive messages, while the method makes no assumptions on agents' homogeneity/interchangeability. The major drawback is that, similarly to CommNet, all agents communicate with everyone. Later works [42] show that the ability of BiCNet to learn effective policies is reduced as the number of the agents increases. This deterioration of effectiveness is attributed to the lack of a mechanism capable of capturing the importance of information from different agents.

TarMAC [43] proposes an architecture designed specifically to allow each agent to chose to which other agents to address messages to, as well as to which messages it will pay attention to. It introduces a signature-based soft attention mechanism with a key, which encodes properties of intended recipients (as opposed to specific agent identification), to be part of the message. The receiver of the message takes this key into consideration and decides whether it is relevant. The method is enhanced by multiple rounds of communication and collaborative reasoning. TarMAC utilizes the actor-critic framework, with a shared policy network and a centralized critic. It is evaluated on four diverse environments against CommNet [38], as well with variants of itself without the attention mechanism or communication at all. The sophistication of the communication scheme could be a disadvantage regarding scalability, as multiple rounds of communication between thousands of agents would result in significant overhead. The main drawback of the method though, with scalability in mind, is the usage of a centralized critic, which takes as input the predicted actions and hidden states of all agents.

Coder [44] is a hierarchical approach, with three distinct hierarchical levels to solve a traffic congestion problem with agents being the traffic intersections. First, a centralized base environment is trained, which is a small and simple subproblem compared with the original one, i.e., a single intersection managing incoming traffic. Here, two training alternatives are considered, a DQN variant and a DDPG variant called Wolpertinger architecture [45]. The next step is called regional DRL, where the parameters of the base environment are shared to a small number of neighboring agents controlling similar but not identical parts of the environment (e.g., different traffic dynamics). To tackle these differences, additional refinement through training is required. The final level combines all regional policies and adds a global dense layer. In order to choose actions while incorporating some form of coordination between the regional policies, an iterative action search is employed. This search starts with the concatenation of the actions produced by the regional nets and attempts to find a globally better alternative. This approach is shown to work with a maximum size of 96 adjacent intersections. A similar approach to Coder [44], we call it here KT DQN [46] expands DQN. Aiming to speed up training and produce better results, it uses single agent training before applying the policy to MAS settings. In doing so, the method freezes all the weights of the single-agent policy network model that are transferred to the MAS scenario, with the exception of the ones between the last hidden layer and the output. Coder and KT DQN utilize a hierarchical approach that initializes learning from a simplified single agent environment. This can indeed speed-up learning, in contrast to start learning from scratch. However, this approach is not straightforwardly transferable between applications. For example, using the regional DRL step of coder calls for a separate design, i.e., decomposing the problem into subproblems, for different applications.

An interesting take on how to utilize an agent population in order to facilitate exploration is presented by J. Leibo [47]. The method proposed is a variant of IMPALA [48]: In the simulation used every agent is a member of an animal species. These homogeneous agents share the same policy network. The paper does not address interagent communication or cooperation, explicitly. Instead, it utilizes the multiagent framework mainly to encourage a more robust exploration. Cooperative behavior emerged when incentives are given to the agents: For example, experiments are presented with agents rewarded for specializing in eating a specific kind of food. The total number of agents present in the simulation reported are 960, and the population is considered to be dynamic.

Mean field MARL [49] proposes two algorithms with the explicit aim to improve MARL scalability. The main idea is to calculate the mean action of an agent's neighborhood, considering that each agent interacts with a virtual mean agent instead of n agents. The two proposed alternatives are mean field Q (MF-Q) and mean field actor-critic (MF-AC), among which the MF-Q approach has superior sample efficiency, which becomes more apparent as the number of agents gets bigger. MF-Q has been empirically shown to scale up to 1000 agents. Mean field MARL does not explicitly address credit assignment, but provides rigorous mathematical proof of convergence to Nash equilibrium. Although both algorithms are scalable, their main drawback lies in the coordination scheme: Averaging the neighborhood of an agent can result in loss of important information and lackluster cooperation [50].

Inspired by the factorization machines [51, 52], FQL [53] utilizes a composite deep neural network architecture, combining Q and V nets, for computing a low-rank approximation of the Q-function, by reformulating the multiagent joint-action Q-function into a factorized form. Depending on agent roles, agents are divided into G groups, and agents within each group share network parameters. While ensuring efficiency, the uniformity assumption between agents might not be suitable for various real-world applications. In addition, since the factorized Q-function for each agent requires the knowledge of the other agents' current states and their last actions for both training and execution, the algorithm mainly addresses the MARL problems with a central controller that communicates the global information to all the agents. In the experimental section, the method produces competitive results in settings with agent populations as large as 500 agents. As the size of the agent population is increased, from 100 to 500, MF-Q [49], which is one of the baselines used, seems to be a more suitable method.

CoLight [54] aims to enable agent cooperation in a large-scale road network with hundreds of traffic signals, recognizing that RL-based methods at the time fail to reach optimal interagent communication. To achieve this, authors introduce the utilization of graph attentional networks. At the core of the method is a Q-value prediction deep network. This network incorporates an observation embedding layer, the hidden neighborhood cooperation layers, and the output Q-value prediction layer. Similarly to mean field MARL [49], the method "averages" neighbors' influence in the broadcasted messages. Although experiments with simulated as well as real-world data involve up to 196 agents, authors claim scalability to thousands of agents, based on method's complexity analysis [54]. Contrasted against various baselines, CoLight shows advanced scalability, as well as sample efficiency.

As already pointed out, many major approaches such as DIAL, CommNet, and BicNet [37, 38, 40] suffer from the lack of agents' ability to differentiate between useful information driving interagent cooperation from noisy or misleading information. ATOC [42] strives to eliminate this specific issue by proposing an attentional communication model that achieves interagent communication between a large amount of agents. The method expands DDPG and uses an attention module as well as a bidirectional LSTM, which serves as a communication channel. The LSTM module plays the role of integrating neighboring agents' internal states, thus creating a message from their combined intentions, and guiding the agents toward coordinated decision-making. Actor and critic networks share parameters to ensure scalability. Indeed, the method seems to challenge the known approaches (DDPG [14], CommNet [38], and BicNet [40]), while agents show division of work and meaningful cooperation.

Graph convolutional reinforcement learning (DGN) [50] has the explicit goal to handle highly dynamic environments, where agents constantly change neighborhoods. Toward this goal, the paper proposes an MAS framework in which agents are connected in a graph where each agent is a node and edges connect neighbors. Neighborhoods are determined by agent distance or other measures, e.g., communication range or critical interactions, and can vary over time. Communication is allowed only between neighbors, in order to minimize inefficiency. DGN consists of an observation encoder, convolutional layers with relation kernels, and a Q network. The method was compared against well-known algorithms such as CommNet [38] and MFQ [49] and is shown to achieve very competitive results in environments with up to 140 agents. However, experiments with greater agent population are needed in order to assess the effectiveness of the method as the environment gets more complex.

Lin et al. [55] proposes two distinct methods, which, very closely to our aims, aim to solve large-scale demand-capacity imbalance problems in the air traffic management domain. The environment simulates a population of approx. 5000 homogeneous agents, acquiring identical rewards. Both methods assume that agents have the same action values. In practice, this assumption means that the agents are not simply homogeneous, but interchangeable. This assumption allows the building of a centralized action-value table, which is utilized for coordination of agents' actions. The first method is called contextual deep Q-learning. It utilizes a table with centralized action values to form a collaborative context, which prevents agents from choosing suboptimal actions, thus avoiding unwanted or redundant behavior such as agents exchanging positions with one another. The second proposed method, called contextual actor-critic, describes an actor critic variant of the contextual DQN. It uses a centralized value function shared by all agents and a parameter-sharing policy network.

Closely related works, with scalability in mind, are those presented by Nguyen et al. [56–58]. In particular, AC for CDec-POMDPs [57] is based on the FEM algorithm [56] and presents an actor-critic algorithm for optimizing collective decentralized POMDPs. It achieves extreme scalability and provides experiments with up to 8000 agents. Subsequent work proposes MCAC [58], which focuses on difference rewards, also addressing the credit assignment problem. A fundamental idea underlining the work described in these papers and a vital aspect in order to achieve scalability is exploiting the count of agents taking the same action a in a state s. This count, as well as other, more complex measures based on this, serves a statistical basis for training, eliminating the need to collect trajectory samples from every agent. Instead, the resulting policy is dependent on count-based observations. Therefore, this method does not explicitly assume communication. It rather assumes full access to all the count-based information during training. During execution, agents execute individual policies without accessing centralized functions. Similarly to contextual DQN [55],

Method	Q Net (s)	V Net	Agents	Communication	MAS population
DQN [3]	I	—	Het	—	2
PS DQN [18]	PS	—	Hom	Impl, global (policy)	200
MADDPG [16]	I	C	Het	Impl, global (critic)	6
VDN [23]	I	—	Het	Impl, global (policy)	2
QMIX [17]	I	—	Het	Impl, global (mixing net)	8
QTRAN [24]	I	C	Het	Impl, global (mixing net)	4
NDQ [25]	I	C	Het	Expl, local (message encoder)	10
ROMA [26]	I	C	Het	Impl, global (mixing net)	27
QPLEX [34]	I	C	Het	Impl, global	27
COMA [36]	I	C	Het	Impl, global (critic)	5
DIAL [37]	I	—	Het	Expl, global (messages)	4
CommNet [38]	C	—	Hom	Expl, global (communication vectors)	500
VAIN [39]	C	—	Hom	Expl, local (attention mechanism)	50
BiCNet [40]	PS	PS	Het	Expl, global (messages in RNN)	32
TarMAC [43]	PS	C	Hom	Expl, local (soft attention mechanism)	20
Coder [44]	PS (KT)	PS (KT)	Hom	Impl, global (policy)	96
KT DQN [46]	PS (KT)	—	Hom	Impl, global (policy)	20
MF-Q [49]	PS	—	Hom	Expl, local (mean agent)	1000
MF-AC [49]	PS	PS	Hom	Expl, local (mean agent)	1000
FQL [53]	PS	PS	Het	Impl, global (policy)	500
CoLight [54]	PS	—	Hom	Expl, local (attention mechanism)	196
ATOC [42]	PS	PS	Hom	Expl, local (attention mechanism/ LSTM)	100
DGN [50]	PS	—	Hom	Expl, local (observations in conv layers)	140
cDQN [55]	PS	—	IC	Expl, local (collaborative context)	5000
cA2C [55]	PS	C	IC	Expl, local (collaborative context)	5000
FEM [56]	PS	PS	IC	Impl, global (policy)	20
AC CDec-POMDPs [57]	PS	PS	IC	Impl, Global (policy)	8000
MCAC [58]	PS	PS	IC	Impl, Global (policy)	8000

Table 1.
Reviewed methods' characteristics.

this approach assumes that the agents are interchangeable. This assumption is necessary to achieve the required scalability, but is not applicable to every problem.

4. Concluding remarks

In this paper, we provide a review of state of the art DMARL methods with significant scalability potential and interagent coordination capabilities in large-scale MAS settings. **Table 1** lists all the reviewed methods with the characteristics, which are considered essential for their scalability, as mentioned in the introductory section. The abbreviations used are the following: Independent learners (I), Parameter Sharing (PS), Centralized model (C), Knowledge Transfer (KT), Hetero/Homo-geneous agents (Het/Hom), Interchangeable agents (IC), Implicit (Impl), Explicit (Expl).

The first conclusion we can draw is the importance of parameter sharing in large agent populations. As the number of agents grows, parameter sharing shows by far the most potential for scalability. It is impractical to train thousands of independent networks for each agent or to utilize a centralized approach whose input size would explode as the number of agents and the size of their observations grow larger. We can clearly see in **Table 1** that all works that provide experiments with large agent populations concur on that approach.

Another important conclusion from this study is the fact that, as scalability potential gets more prominent, stricter assumptions are made on the agents and on their environment. Specifically, the approach that manages to scale up to the largest agent population [57] assumes interchangeable agents. In order for these methods to find broad practical applicability, such assumptions have to be relaxed. In numerous practical applications, agents are not homogeneous, but more importantly, interchangeable agents are a more rare occurrence. Additions to the methods, in order to incorporate heterogeneous agents, would be of great value. An important point here is that parameter sharing assumes homogeneous agents. Further research and experimenting on the direction of techniques such as labeling agents, as proposed in the original parameter sharing publication [18], should be beneficial in that regard.

Finally, we can conclude that as the need for scalability becomes more prevalent, targeted and detailed communication becomes more challenging to achieve. For example, MF-Q [49], a method that shows great scalability capabilities, assumes that all agents affect the recipient of their messages equally. Local communication, on the other hand, affects the scalability of methods, while further studies on the use of attention and messages' combination mechanisms are necessary to prove the potential to operate in environments with thousands of agents.

Acknowledgements

This work is supported by the TAPAS project, which has received funding from the SESAR Joint Undertaking (JU) under grant agreement No 892358. In addition, this work was partially supported by the University of Piraeus Research Center (UPRC).

Author details

Theocharis Kravaris* and George A. Vouros
University of Piraeus, Piraeus, Greece

*Address all correspondence to: hariskrav.unipi@gmail.com

References

[1] Kravaris T, Spatharis C, Bastas A, Vouros GA, Blekas K, Andrienko G, Andrienko N, Garcia JM. Resolving congestions in the air traffic management domain via multiagent reinforcement learning methods. 14 December 2019. arXiv preprint arXiv: 1912.06860

[2] Sutton RS, Barto AG. Reinforcement learning: An introduction. Cambridge, Massachusetts, USA: MIT Press; 13 November, 2018

[3] Mnih V, Kavukcuoglu K, Silver D, Rusu AA, Veness J, Bellemare MG, et al. Human-level control through deep reinforcement learning. Nature. 2015; **518**(7540):529-533

[4] Tsitsiklis J, Van Roy B. Analysis of temporal-diffference learning with function approximation. Advances in Neural Information Processing Systems. 1996;**9**:1075-1081

[5] Wang Z, Schaul T, Hessel M, Hasselt H, Lanctot M, Freitas N. Dueling network architectures for deep reinforcement learning. In: International Conference on Machine Learning. New York, NY, USA: JMLR: WCP; 11 June 2016. pp. 1995-2003

[6] Fortunato M, Azar MG, Piot B, Menick J, Osband I, Graves A, et al. Noisy networks for exploration. 30 June 2017. arXiv preprint arXiv:1706.10295

[7] Mnih V, Badia AP, Mirza M, Graves A, Lillicrap T, Harley T, et al. Asynchronous methods for deep reinforcement learning. In: International Conference on Machine Learning. New York, NY, USA: JMLR: W&CP; 2016. pp. 1928-1937

[8] Dabney W, Rowland M, Bellemare M, Munos R. Distributional reinforcement learning with quantile regression. In: Proceedings of the AAAI Conference on Artificial Intelligence. California, USA: AAAI Press, Palo Alto; 2018

[9] Hessel M, Modayil J, Van Hasselt H, Schaul T, Ostrovski G, Dabney W, et al. Rainbow: Combining improvements in deep reinforcement learning. In: Thirty-second AAAI Conference on Artificial Intelligence. California, USA: AAAI Press, Palo Alto; 2018

[10] Van Hasselt H, Guez A, Silver D. Deep reinforcement learning with double q-learning. In: Thirty-second AAAI Conference on Artificial Intelligence. California, USA: AAAI Press, Palo Alto; 2016

[11] Schaul T, Quan J, Antonoglou I, Silver D. Prioritized Experience Replay. InICLR (Poster). 1 January 2016

[12] Hasselt H. Double Q-learning. Advances in Neural Information Processing Systems. 2010;**23**:2613-2621

[13] Tampuu A, Matiisen T, Kodelja D, Kuzovkin I, Korjus K, Aru J, et al. Multiagent cooperation and competition with deep reinforcement learning. PloS One. 2017;**12**(4):e0172395

[14] Qiu C, Hu Y, Chen Y, Zeng B. Deep deterministic policy gradient (DDPG)-based energy harvesting wireless communications. IEEE Internet of Things Journal. 2019;**6**(5): 8577-8588

[15] Silver D, Lever G, Heess N, Degris T, Wierstra D, Riedmiller M. Deterministic policy gradient algorithms. In: International Conference on Machine Learning. New York, NY, USA: JMLR: W&CP; 2014. pp. 387-395

[16] Lowe R, Wu YI, Tamar A, Harb J, Pieter Abbeel O, Mordatch I. Multi-agent actor-critic for mixed cooperative-competitive environments. Advances in Neural Information Processing Systems. 2017;**30**:6379-6390

[17] Rashid T, Samvelyan M, Schroeder C, Farquhar G, Foerster J, Whiteson S. Qmix: Monotonic value function factorisation for deep multi-agent reinforcement learning. In: International Conference on Machine Learning. New York, NY, USA: JMLR: W&CP; 2018. pp. 4295-4304

[18] Gupta JK, Egorov M, Kochenderfer M. Cooperative multi-agent control using deep reinforcement learning. In: International Conference on Autonomous Agents and Multiagent Systems. Cham, Switzerland: Springer International Publishing AG; 2017. pp. 66-83

[19] Castaneda AO. Deep Reinforcement Learning Variants of Multi-agent Learning Algorithms. Edinburgh: School of Informatics, University of Edinburgh; 2016

[20] Agogino AK, Tumer K. Analyzing and visualizing multiagent rewards in dynamic and stochastic domains. Autonomous Agents and Multi-Agent Systems. 2008;**17**(2):320-338

[21] Devlin SM, Kudenko D. Dynamic potential-based reward shaping. In: Proceedings of the 11th International Conference on Autonomous Agents and Multiagent Systems. Cham, Switzerland: Springer International Publishing AG; 2012. pp. 433-440

[22] Ng AY, Harada D, Russell S. Policy invariance under reward transformations: Theory and application to reward shaping. In: ICML. New York, NY, USA: JMLR: W&CP; 1999. pp. 278-287

[23] Sunehag P, Lever G, Gruslys A, Czarnecki WM, Zambaldi VF, Jaderberg M, et al. Value-decomposition networks for cooperative multi-agent learning based on team reward. In: AAMAS. Cham, Switzerland: Springer International Publishing AG; 2018

[24] Son K, Kim D, Kang WJ, Hostallero DE, Yi Y. Qtran: Learning to factorize with transformation for cooperative multi-agent reinforcement learning. In: International Conference on Machine Learning. New York, NY, USA: JMLR: W&CP; 2019. pp. 5887-5896

[25] Wang T, Wang J, Zheng C, Zhang C. Learning nearly decomposable value functions via communication minimization. In: International Conference on Learning Representations. 2019

[26] Wang T, Dong H, Lesser V, Zhang C. ROMA: Multi-agent reinforcement learning with emergent roles. In: International Conference on Machine Learning. New York, NY, USA: JMLR: W&CP; 2020. pp. 9876-9886

[27] Becht M, Gurzki T, Klarmann J, Muscholl M. ROPE: Role oriented programming environment for multiagent systems. In: Proceedings Fourth IFCIS International Conference on Cooperative Information Systems. CoopIS 99 (Cat. No. PR00384). New York, U.S.: IEEE; 1999. pp. 325-333

[28] Stone P, Veloso M. Task decomposition, dynamic role assignment, and low-bandwidth communication for real-time strategic teamwork. Artificial Intelligence. 1999; **110**(2):241-273

[29] Depke R, Heckel R, Küster JM. Roles in agent-oriented modeling. International Journal of Software Engineering and Knowledge Engineering. 2001;**11**(03):281-302

[30] Ferber J, Gutknecht O, Michel F. From agents to organizations: An organizational view of multi-agent systems. In: International Workshop on Agent-oriented Software Engineering. Cham, Switzerland: Springer International Publishing AG; 2003. pp. 214-230

[31] Odell J, Nodine M, Levy R. A metamodel for agents, roles, and groups. In: International Workshop on Agent-oriented Software Engineering. Cham, Switzerland: Springer International Publishing AG; 2004. pp. 78-92

[32] Cossentino M, Hilaire V, Molesini A, Seidita V, editors. Handbook on Agent-oriented Design Processes. Berlin Heidelberg: Springer; 2014

[33] Lhaksmana KM, Murakami Y, Ishida T. Role-based modeling for designing agent behavior in self-organizing multi-agent systems. International Journal of Software Engineering and Knowledge Engineering. 2018;**28**:79-96

[34] Wang J, Ren Z, Liu T, Yu Y, Zhang C. QPLEX: Duplex dueling multi-agent Q-learning. In: International Conference on Learning Representations. 2020

[35] Samvelyan M, Rashid T, de Witt C, Farquhar G, Nardelli N, Rudner TG, et al. The starcraft multi-agent challenge. In: Proceedings of the 18th International Conference on Autonomous Agents and MultiAgent Systems. Cham, Switzerland: Springer International Publishing AG; 2019. pp. 2186-2188

[36] Foerster J, Farquhar G, Afouras T, Nardelli N, Whiteson S. Counterfactual multi-agent policy gradients. In: Proceedings of the AAAI Conference on Artificial Intelligence. California, USA: AAAI Press, Palo Alto; 2018

[37] Foerster J, Assael IA, De Freitas N, Whiteson S. Learning to communicate with deep multi-agent reinforcement learning. Advances in Neural Information Processing Systems. 2016;**29**:2137-2145

[38] Sukhbaatar S, Fergus R. Learning multiagent communication with backpropagation. Advances in Neural Information Processing Systems. 2016;**29**: 2244-2252

[39] Hoshen Y. Vain: Attentional multi-agent predictive modeling. Advances in Neural Information Processing Systems. 2017;**30**:2701-2711

[40] Peng P, Wen Y, Yang Y, Yuan Q, Tang Z, Long H, Wang J. Multiagent bidirectionally-coordinated nets: Emergence of human-level coordination in learning to play starcraft combat games. 29 March 2017. arXiv preprint arXiv: 1703.10069

[41] Schuster M, Paliwal KK. Bidirectional recurrent neural networks. IEEE Transactions on Signal Processing. 1997;**45**(11):2673-2681

[42] Jiang J, Lu Z. Learning attentional communication for multi-agent cooperation. Advances in Neural Information Processing Systems. 2018; **31**:7254-7264

[43] Das A, Gervet T, Romoff J, Batra D, Parikh D, Rabbat M, et al. Tarmac: Targeted multi-agent communication. In: International Conference on Machine Learning. New York, NY, USA: JMLR: W&CP; 2019. pp. 1538-1546

[44] Tan T, Bao F, Deng Y, Jin A, Dai Q, Wang J. Cooperative deep reinforcement learning for large-scale traffic grid signal control. IEEE Transactions on Cybernetics. 2019;**50**(6):2687-2700

[45] Marden JR, Arslan G, Shamma JS. Cooperative control and potential games.

IEEE Transactions on Systems, Man, and Cybernetics, Part B (Cybernetics). 2009; **39**(6):1393-1407

[46] Rădulescu R, Legrand M, Efthymiadis K, Roijers DM, Nowé A. Deep multi-agent reinforcement learning in a homogeneous open population. In: Benelux Conference on Artificial Intelligence. New York, NY, USA: JMLR: W&CP; 2018. pp. 90-105

[47] Leibo JZ, Pérolat J, Hughes E, Wheelwright S, Marblestone AH, Duéñez-Guzmán EA, et al. Malthusian Reinforcement Learning. In: AAMAS. Cham, Switzerland: Springer International Publishing AG; 2019

[48] Espeholt L, Soyer H, Munos R, Simonyan K, Mnih V, Ward T, et al. IMPALA: Scalable distributed deep-RL with importance weighted actor-learner architectures. In: ICML. New York, NY, USA: JMLR: W&CP; 2018

[49] Yang Y, Luo R, Li M, Zhou M, Zhang W, Wang J. Mean field multi-agent reinforcement learning. In: International Conference on Machine Learning. New York, NY, USA: JMLR: W&CP; 2018. pp. 5571-5580

[50] You J, Liu B, Ying Z, Pande V, Leskovec J. Graph convolutional policy network for goal-directed molecular graph generation. Advances in Neural Information Processing Systems. 2018; **31**:6410-6421

[51] Rendle S, Schmidt-Thieme L. Pairwise interaction tensor factorization for personalized tag recommendation. In: Proceedings of the Third ACM International Conference on Web Search and Data Mining. New York City, New York, USA: ACM. 2010; pp. 81-90

[52] Rendle S. Factorization machines with libfm. ACM Transactions on Intelligent Systems and Technology (TIST). 2012;**3**(3):1-22

[53] Zhou M, Chen Y, Wen Y, Yang Y, Su Y, Zhang W, et al. Factorized q-learning for large-scale multi-agent systems. In: Proceedings of the First International Conference on Distributed Artificial Intelligence. New York City, New York, USA: ACM; 2019. pp. 1-7

[54] Wei H, Xu N, Zhang H, Zheng G, Zang X, Chen C, et al. Colight: Learning network-level cooperation for traffic signal control. In: Proceedings of the 28th ACM International Conference on Information and Knowledge Management. New York, NY, United States: Association for Computing Machinery; 2019. pp. 1913-1922

[55] Lin K, Zhao R, Xu Z, Zhou J. Efficient large-scale fleet management via multi-agent deep reinforcement learning. In: Proceedings of the 24th ACM SIGKDD International Conference on Knowledge Discovery & Data Mining. New York, NY, United States: Association for Computing Machinery; 2018. pp. 1774-1783

[56] Nguyen DT, Kumar A, Lau HC. Collective multiagent sequential decision making under uncertainty. In: Thirty-First AAAI Conference on Artificial Intelligence. California, USA: AAAI Press, Palo Alto; 2017

[57] Nguyen DT, Kumar A, Lau HC. Policy gradient with value function approximation for collective multiagent planning. Advances in Neural Information Processing Systems. 2017; **30**:4319-4329

[58] Nguyen DT, Kumar A, Lau HC. Credit assignment for collective multiagent RL with global rewards. Advances in Neural Information Processing Systems. 2018;**31**:8102-8113

Section 2

Applications of Multi-Agent Technologies Combined with Machine Learning

Chapter 3

Role of an Optimal Multiagent Scheduling in Different Applications Using ML

Fahmina Taranum, Sridevi K, Maniza Hijab, Syeda Fouzia Sayeedunissa, Afshan Kaleem and Niraja K.S

Abstract

Scheduling is regarded as one of the vital decision-making processes used frequently in many real-time cases. It manages everything from resource allocation to the task completion, with the goal to optimize the desired objectives. Subject to the problem, the resources, tasks, and goals can differ. The aim is to design a corporative multiagent system for optimal scheduling. Many of the scheduling available algorithms calculate optimality based on different perspectives. The proposal is to create the dataset using multiple algorithms with different performance metrics to find an optimal one. This data can be imported into machine learning tools for training and predicting, based on the selected performance metrics. The algorithm considered in the empirical analysis includes first come first serve, Round robin, and Ant colony approach. The major finding shows that scheduling using Ant colony is an optimal algorithm, which is based on speed and velocity. The future extension would be to check the correctness of optimality using machine learning tools.

Keywords: optimality, scheduling, machine learning

1. Introduction

1.1 Scheduling

The scheduling task can be comprehended as a distributed consecutive decision-making process, which is designed using multiagent reinforcement learning algorithms. These algorithms provide the agents with effective learning as they involve frequent interactions with the environment, thereby enhancing their relevance to numerous real-time cases. The scheduling entities used are resources and tasks, which are interchangeably assigned to each other. The resources can be classified as heterogeneous and homogeneous. Based on the generation of throughput, resources are classified as homogenous resources with similar or uniform throughput and heterogeneous refers to distinct type. Dependent tasks are interconnected job endeavors, in which the task cannot initiate until the accomplishment of a separate task. The characteristics based on defining the task include priority and QoS. Another widely used scheduling mechanism is based on the notion of priority rules. In this method, the resource allocation and task execution are scheduled by designating some priorities to them as per the situation, demand, requirement, or need.

The priority can be based on the deadline and load. The magnitude of work carried out to processes for manufacturing components is termed as load, which must be balanced to execute fair scheduling. Deadlines are the limits or constraints to complete the job, which can be classified as firm, soft, and hard. Based on the applicability the classification is named as soft if is not rigid, or if some limited consideration is given then it is firm, or if it is not accepted after the time limit then hard The components used for the entities are depicted in **Figure 1**.

1.2 Optimal schedule

Optimality can be measured with respect to task or resource using multiple metrics, as shown in **Figure 2**. Most of the optimization practices are extremely nonlinear and cross-media under different contexts and constraints with high convergence performance and low computational cost. It can be classified as gradient or derivative, stochastic or deterministic, or population-based or trajectory-based.

The word optimal means the best and most desirable solution and scheduling means arranging, controlling, and optimizing work and workloads. From that sequence of job, there should be maximum utilization of machines and less waiting time for the job and maximum work should be done by satisfying various constraints. It is an optimization problem in scheduling where the input to the machine is the list of jobs and the output is the schedule of all the jobs on the particular machine. The schedule should optimize the throughput and resource utilization. This is also known as machine scheduling, processor scheduling, multiprocessor scheduling, or just scheduling. The jobs can be scheduled on single machine or multiple machines based on the requirements. Either a set of jobs given to only one machine or parallel to more than one machine.

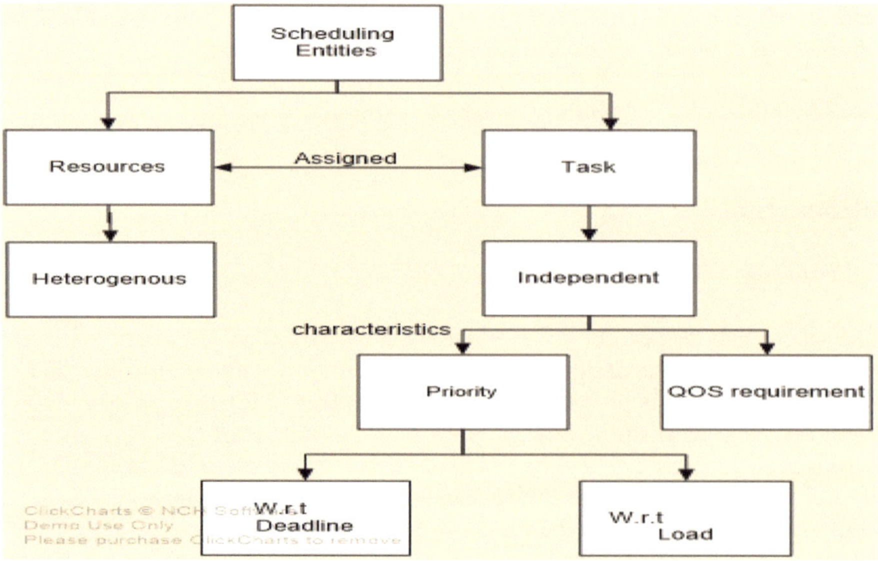

Figure 1.
Scheduling entities.

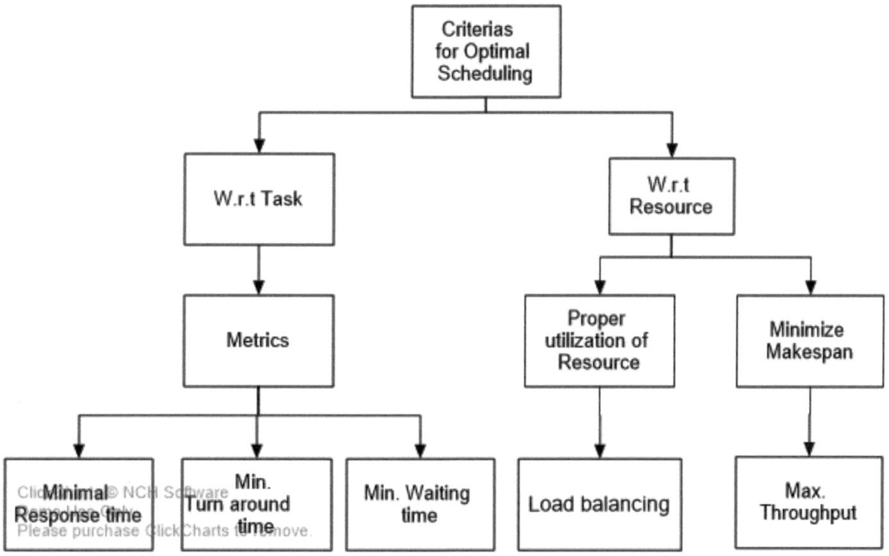

Figure 2.
Criteria for optimal scheduling.

The criteria for scheduling a resource are its proper utilization, generating minimum makespan, and maximum throughput. Utilization refers to the instance of manufacturing a component based on its real-world or commercial usage. Makespan refers to the time taken by the units of resources to properly complete executing all the jobs. Throughput refers to the process of executing the job in a unit amount of time.

There are various parameters considered to get the optimal job or task for setup.

- Start dates and deadlines of the job
- Costs depend on the completion times of activities
- Possibilities of leaving some jobs "unperformed." due to some interrupt.
- Initial Setup times and costs.

If there is a machine M with a set of jobs J1, J2, and J3 that needs to be executed by M, with maximum work in minimum time to complete all jobs with the best sequence is called optimal job scheduling.

1.3 Types of scheduling

There are many algorithms for scheduling of dependent or independent tasks in a single processor or multiprocessor environment with different speeds. And using dynamic programming algorithms, the best optimal scheduling can be obtained by considering the time taken to complete the job or priority of the job and any other constraints need to satisfy to finish the particular tasks.

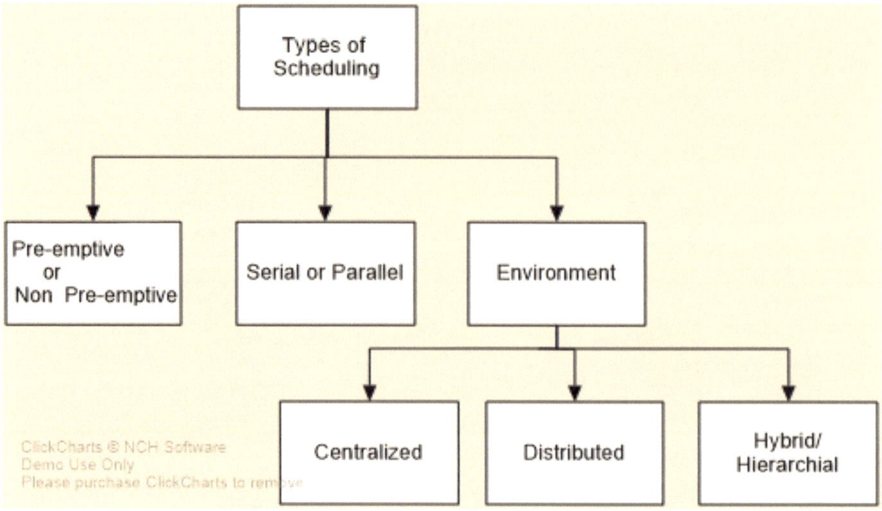

Figure 3.
Types of scheduling.

The effective allocation of the order of the jobs to a machine to get the maximum profit and minimum cost from the process leads to optimal scheduling is depicted in **Figure 3**.

Classification of the scheduling can be preemptive or nonprimitive, in which the decision to take the resource from the low-priority process and allocate it to a higher priory one is taken. For the simultaneous, linear or nonlinear class of execution, the second level of classification is used. Finally, the decision to distribute the resource can be done at central, distributed, or combined by using different environments based on the demand and architecture used. The most widely used scheduling algorithms are Round-robin, First come first serve, Shortest job first, Earliest Deadline First, Priority Scheduling, Multi-Level Queue Algorithm, Multi-level Feedback Queue Scheduling Algorithm, and shortest Remaining Time.

1.4 Approaches used in machine learning

Reinforcement learning, which is usually preferred is an area of machine learning which emphasizes how intelligently an agent decides an environment to achieve the optimum reward. This mechanism allows learning using trial-and-error communication with the environment. Scheduling helps to interpret the data accurately, if collaborated with machine learning as it generates the optimal results based on predictions as depicted in **Figure 4**. Through frequent trials, the agents come upon different possible outcomes for an action and thereby find the most appropriate action to be done in any given situation.

For instance, upon encountering an unexpected situation, reinforcement learning can be found helpful, as it enables learning from prior results and altering the parameters as per the need before passing for the subsequent iterations thereby ensuring that the solutions are as desired and robust.

The learning algorithms could benefit from this idea, by associating the priorities with the feedback signal the agents receive when executing the actions. Later by incorporating the priority rules with the feedback received for each action, the learning algorithm can be improvised. The aim is to generate multiple cases from different

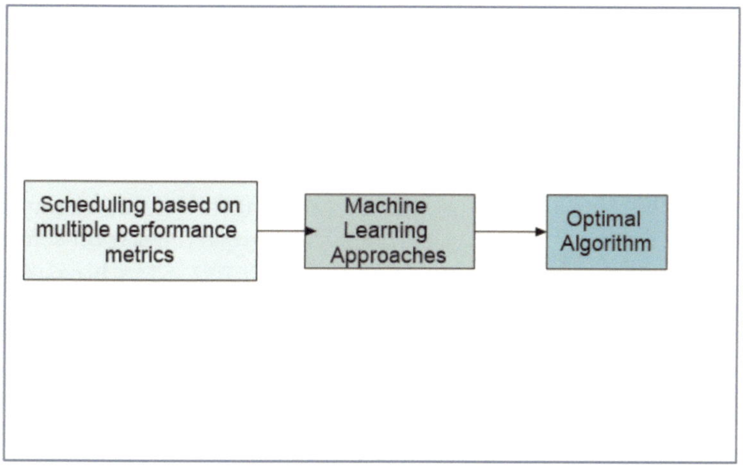

Figure 4.
Machine learning and optimality.

algorithms and train a machine with the data generated from execution, to validate the optimality of scheduling using machine-generated test results.

1.5 Applications

Based on the usage of multitask applications in every field, a need for efficient utilization of optimal multiagent scheduling algorithms is highlighted. Machine learning algorithms are widely used to improve the optimality of multiagent scheduling in the field of production and transportation. The major domains where some of these challenges can be improved have been discussed below.

1.5.1 Transportation domain

Every individual had to rely on some mode of transportation which can be through land, air, or water. With the advancement of technology, the usage of air and land transportation has drastically improved due to efficient time management and comfort. The development of many multiagent system software has been of the major reasons for this. Though the multiagent scheduling softwares still exist, there are some challenges that need to be improved.

1.5.2 Railway domain

Transportation is the major source of people especially above 75% of people relies on railway networks throughout the globe. When it comes to railway networks, most railways run on a single line where more limitations exist for decades. Although multiagent systems have applied to this, many of the existing challenges are unavoidable. The train delay rate is a challenging issue to date, even with the advances in technology, it cannot be handled perfectly. With the advancement of machine learning algorithms, we can find a path with the optimal multiagent scheduling algorithm. Electricity transport management plays a vital role, especially in railway networks with the utilization of machine learning algorithms. We can find an optimal solution for scheduling the multiagent system to choose the proper routing to overcome the major challenges faced by the transportation domain.

1.5.3 Airline reservation domain

Airline flight bidding software mostly uses multiagent system prototype software, where the buyers buy the e-ticket. The optimal multiagent scheduling algorithm with machine learning can direct the buyer to a better price line system through which buyers take preferences and correlate the parameters with the available flights.

1.5.4 Manufacturing domain

Every business domain has its own production department where different jobs have to be scheduled with multiagent approaches. Machine Learning usage has become unimaginable heights in the last few years where the manufacturing industry's economic growth rate can be improved with the utilization of optimal multiagent scheduling algorithms.

1.5.5 Electronic commerce domain

This domain will show its increased growth rate in the near future as most of the day-to-day transactions are carried out these days through e-commerce combined with the multiagent scheduler and machine learning algorithms.

2. Literature review

The proposal by Kumar et al. [1] is to enterprise a scheme for processing tasks to allocate resources at runtime. The experimentation was performed on Cloudsim to validate the effective technique for optimum solutions. The proposal aims at tracing results with initial high velocity, which is gradually decreased improved exploitation.

The proposal of reinforcement learning is used by author Chi Zhang et al. [2], to achieve corporation in the scheduling of multiagent working in a team, which is an agent feedback-based learning used to learn from the results of the environment and perform action. The concept of Markov chain and Proximal Policy Optimization is used to check for optimal scheduling. The performance metrics used are load consumption and price of electricity to check for optimal results. The optimality of the results is based on the trained dataset or historical data used. Multiagent reinforcement learning is used when manifold inhabitants with multiple global constraints interact with different scheduling errands. This approach is used in a realistic environment to obtain optical results using predictions.

The population diversity control problem is selected in the proposal by author Zha oyun Song, Bo Liu et al. [3] to improve optimization. The method is based on the runtime selection for adaptive and diversified controlling parameters. The accuracy is calculated for the cluster of movable particles.

The concept of using low voltage over harmonic distortion of modulated index is used to set the image threshold for increasing the image accuracy by the author [4].

The scheduling of multiagent in a cooperative, spatial and temporal constraints is used by the author Julie Shah [5] to obtain an optimal task assignment. The experimental analysis is done on the hill climbing algorithm using intricate computation and the accuracy versus optimality is compared with the conventional algorithm. An empirical result proves that the optimality of hill climbing is better using the mixed integer collaboration approach of linear programming.

The author Khaled M. Khalil [6] has used the Netlogo simulator for experimentation. The aim is to maximize the agent group by maximizing the reward to the agent working in the group using Q-learning algorithm (action value function) along with working in an interactive, autonomous, and seamless environment. The approaches of high relevance to realistic solutions grounded on AI was proposed by authors Martin Riedmiller et al. [7] to optimize, manufacture and control the scheduling based on distributed sequential approaches for employment related decision making with multi agent reinforcement learning.

The Authors E. Grace Mary Kanaga [8] have used an approach of ANT colony for an optimal scheduling algorithm. The continuous optimization problems are solved using velocity, inertia weight location of the particle, and global and local best positions. Scheduling patients for an optimal accurate solution is performed.

Zhi hui et al. [9] have aimed to implement adaptive behavior for population distribution along with fitness.

The target is to improve and validate the global convergent ability by authors Jianchao Zeng, Jing Jie et al. [10]. Kennedy et al. [11] have invented the perception using non-linear functions of Ant colony using particle swarm methodologies for optimization. Kennedy et al. [11] is the inventor behind the use of non-linear functions to design methodologies based on cluster optimization. The relationships amongst PSO and its integration with artificial life using genetic algorithms are proposed and discussed.

The author Wilfried Brauer et al. [12], aims at scheduling multimachine, and assigning jobs based on demand like cost, the effectiveness of results, and time. The approach of artificial intelligence is used to collect and train the data in distributed, parallel, asynchronous and corporative multi-agent environments. To generate the results two learning steps known as successor selection and estimate adjustment are repeatedly applied and experimented on individual machines.

3. Proposal

The idea is to generate optimal scheduling, which is tested for set of scheduling algorithms and the major finding is ANT colony scheduling gave the best results. The algorithm aims to predict and generate the best position using parameters like inertia, weight, speed, velocity, distance, acceleration constants, direction, and local and global position. The local and global positions are the best positions for each and the group of jobs, respectively.

The movement of the job is in the direction of the best location value. The velocity is calculated using Eq. (1).

$$Ve_i^{z+1} = wVe_i^{z} + ac_1p_1\left(Lbst_i^{z} - x_i^{z}\right) + ac_2p_2\left(Gbst^{k} - x_i^{z}\right) \quad (1)$$

Lbst is the minimum value of the experience position of each job and the group's minimum value of experience position is Gbst. The movement is monitored using the kth step of ith job to get the next new place. The first term of Eq. (1) gives the inertia with respect to previous velocity; second and third terms are used to fetch the direction of each and group of job respectively. An accelerating object (ac) is used to persuade a uniform alteration in its velocity at each instant. Acceleration constants used in Eq. (1) are ac1 and ac2. The random numbers are p1 and p2 [0, 1] of range are chosen. Velocity and Position update is derived using Eq. (1) as given by Kennedy and Eberhart [11].

$$x^{z+1}_j = x^z_j + Ve^{z+1}_j \tag{2}$$

Eq. (2) is used to get the new position from the previous position and the velocity of its movement. After getting the best and optimal position, all the particles synchronize with each other. According to Kennedy et al. [11], the search space is epitomized in a D-dimensional vector and the velocity and position update are calculated using Eqs. (1) and (2). The jth particle's position in D-dimensional vector is calculated using Eq. (3). The local best (Lbst) and the global best (Gbst) of the jth particle are denoted in Eqs. (3) and (4) respectively.

$$x_j = (x_{j1}, x_{j2}, x_{j3}, x_{jk}, \dots, x_{jd}) \tag{3}$$

$$vex_j = (x_{j1}, x_{j2}, x_{j3}, x_{jk}, \dots, x_{jd}) \tag{4}$$

$$l_j = (x_{j1}, x_{j2}, x_{j3}, x_{jk}, \dots, x_{jd}) \tag{5}$$

$$g_j = (x_{1j}, x_{2j}, x_{3j}, x_{ij}, \dots, x_{jd}) \tag{6}$$

For scheduling a process with recourse, the problem is to schedule 'n' jobs with appropriate resources; each process has a set of sequential tasks (operations) and an index of the job. The count of operations used in scheduling is denoted by D in (7).

$$D = \sum n_i \tag{7}$$

$$Y = \sum_{i=1}^{i-1} n_i + 1 \tag{8}$$

3.1 Optimization for job scheduling

To generate an ideal solution for Job-scheduling multiple scheduling algorithms are available, which works on different parameters. To generate an optimal one, the need to select the parameter for performance metric becomes mandatory, based on which the complete experimentation can be carried out. Let the proposal be defined on the continuous optimization parameters like particle position xj, Velocity vej, acceleration coefficients ac1 and ac2, and inertia weight ω. Scheduling a task is a conjunctional and feasible optimization with sequence of selected resource along with its operation. The aim is to find an optimal schedule without busy waiting and with fair scheduling.

3.2 Task scheduling with optimization

The scenario consisting of n task (or jobs) denoted as J = {J1, J2, ..., Jn}, sequential task T={1 ... n}, Resource R = {R_1, R_2, ..., R_m}, and O_{ij} are the operations with i, j are the indices of the jobs and task respectively. Every task of the vth job is numbered as v and for same number of v tasks in **Table 1**, and y is defined in Eq. (8). One sample representation of the job for optimization of scheduling is depicted in **Table 1**.

a. Phase I: Scheduling for three jobs with three resources is depicted in **Table 2**. The experimentation is done in multiple permutations (3p3 ways for 3,3 job, resource allocation) giving in total 9 such task possibility. **Table 2** uses a few of the nine possible permutations, where R1, R2, and R3 are the distinct resources. The example for job representation along with position and velocity vector initialization of scheduling is given in **Table 3**.

   Initialization of the particle position is done with random numbers from x_{min} to x_{max}, and x_{min} is set to 0 and x_{max} is set to 2. The velocity vector is initialized with the random numbers limitation of v_{min} as −4 and v_{max} as 4 as shown in **Table 3**.

b. Phase II: Decoding Particles with job solutions: An integration of optimizing for job scheduling cannot be deployed directly as a solution to particle position. Hence, indirect ways are adopted to decode particle representation as a solution to schedule job problem. For decoding jobs into a schedule, the algorithmic steps are listed below:

Task	1	2	...	V	...	n
Position X_{ij}	Xi1	Xi2	...	Xiy	...	Xid

Table 1.
Job's representation for resource scheduling.

Job	Arrival time	Task Sequence		Processing time		Age
1	1	**R1 R2 R3**	2	1	3	38
2	3	R2 R3 R1	1	3	2	40
3	5	R3 R2 R1	3	1	2	44

Table 2.
Job scheduling problem for 3:3 jobs and resource allocation.

Task	1	1	1	2	2	2	3	3	3
Position xij (Values)	xi1 (.81)	xi2 (1.12)	xi3 (1.86)	xi4 (1.09)	xi5 (.56)	xi6 (0.93)	xi7 (.68)	xi8 (1.95)	xi9 (1.76)
Velocity vij (Values)	vi1 (−2.96)	vi2 (1.67)	vi3 (−2.93)	vi4 (.99)	vi5–3.63	vi6 (.86)	vi7 (−3.10)	vi8 (3.26)	vi9 (.09)

Table 3.
Job representation along with initialization of position and velocity vector for i^{th} job.

Task unit	2.1	3.1	1.1	2.2	2.3	1.2	3.2	1.3	3.3
Position Pij	Xi5	Xi7	Xi1	xi6	Xi4	Xi2	Xi9	xi3	Xi8
(Values)	0.56	0.68	0.81	0.93	1.09	1.12	1.76	1.86	1.95
Order of execution	1	2	3	4	5	6	7	8	9

Table 4.
Result obtained for sequenced particles after phase II.

R3		J 3.1			J 2.2			J 1.3							
R2	J 2.1		J 1.2	J 3.2											
R1	J1.1				J 3.3		J2.3								
Time slot	1	2	3	4	5	6	7	8	9	10	11	12	13	14	15

Table 5.
Decoded schedule with optimization.

Step I: sort in ascending order the values of the position vector.

Step II: arrange the tasks in the corresponding order of the values of the position vector obtained in step I.

Step III: The resultant is with sequential order of task along with the corresponding positions as shown in **Table 4**.

Using the sequence obtained with operation-based permutation is π = (2,3,1,2,2,1,3,1,3). An element of π with a value i for Job Ji. The jth occurrence of i in π refers to operation Oij for the jth task (operation) of Job ith. The precedence of the task is determined simply by the order of the elements of π and all the task are ready for scheduling as per the first row of **Table 4**. Based on the permutation, the first element selected for scheduling according to the permutation is 2, therefore, first unit of the third Job is processed on resource R3.

Followed with the first task of the second Job is processed on R2, and then the first unit of the task1 is scheduled on R1. The obtained decode schedule is shown in **Table 5**.

Jobs 1,2, and 3 complete the execution at 9, 8, and 6 time respectively as concluded from **Table 5** after applying optimization algorithm based on the order of tasks, as depicted below:

$$O = \{O31, O21, O11, O12, O22, O32, O33, O13, O23\}$$

It is optimal scheduling as all the processes have been completed within time limit of 9, with the time of completion for jobs as {J1, J2, J3 as 9,8,6}.

3.3 First come first server algorithm

The working concept of FCFS is to serve the first in jobs on high priority, and results obtained after applying the scheduling are shown in **Table 6**.

R3					J 1.3			J 2.2			J 3.1									
R2			J 1.2	J 2.1										J3.2						
R1	**J 1.1**									**J 2.3**					**J3.3**					
Time slot	1	2	3	4	5	6	7	8	9	10	1 1	1 2	1 3	14	1 5	1 6	1 7	1 8	1 9	2 0

Table 6.
Decoded schedule with FCFS.

R3					J2,2		J3.1			J2.3			J1.3							
R2				J2.1	J1.2					J3.2										
R1		**J1,1**									**J3.3**									
Time slot	1	2	3	4	5	6	7	8	9	10	11	12	13	14	15	16	17	18	19	20

Table 7.
Job decoded schedule using priority.

Jobs 1,2, and 3 complete the execution at 6, 11, and 16 times respectively after applying the FCFS scheduling algorithm as concluded from **Table 6**. The order is based on first-entered jobs and would be the first to be catered.

The selected operation of tasks is as depicted below:

$$O = \{O11, O12, O21, O13, O22, O31, O23, O32, O33\}$$

Hence, the conclusion is drawn as FCFS is not optimal when compared to optimization algorithm. Job 1 has arrived first and he/she is processed for task 1 on resource R1 which is available. Next Job 1 task 2 has to be processed on resource R2 and after completion of the task; Job 1 has to get the resource R3 for task 3. Thus, based on the FCFS description **Table 6** decoded schedules are obtained. Based on FCFS the calculated total job completion time is:

$$\{J1 = 6, J2 = 11 \text{ and } J3 = 16\}$$

3.3.1 Based on age priority (using preemptive manner)

The working concept of priority is based on age priority, and results obtained after applying the scheduling are depicted in **Table 7**.

Job 1 is processed on task 1 with a request for resource R1, Job 2 needs resource R2 for its first task and based on its availability is allocated for 1 slot. Makespan time for jobs is {J1 = 14–1 = 13, J2 = 14–3 = 4, J3 = 12–5 = 7}, i.e{J1, J2, J3 is 13, 4 and 7, respectively}.

4. Conclusion

Multi-agent technologies and approaches can be applied to machine learning, based on their capabilities of flexibility, adaptability and self-sufficiency. The applications designed using these methodologies are realistic, dynamic and distributive in nature. The idea is to build a best decision making model using the latest approaches of machine learning. The experiments are tested for four types of Scheduling Round robin, FCFS, Priority-based and Ant colony or particle swarm optimization technique. The results validate the Ant colony approach with the optimal answer. The dataset generated with this experimentation for multi-agent scheduling is collected with different performance metrics. The future work of this proposal is to justify the validation of multiagent scheduling using a machine learning tool, by training the machine using the generated dataset.

Author details

Fahmina Taranum[1], Sridevi K[1], Maniza Hijab[1]*, Syeda Fouzia Sayeedunissa[1], Afshan Kaleem[1] and Niraja K.S[2]

1 Muffakham Jah College of Engineering and Technology affiliated to Osmania University, Hyderabad, Telangana, India

2 BVRIT Hyderabad College of Engineering for Women, Hyderabad, Telangana, India

*Address all correspondence to: hijabmaniza@gmail.com

References

[1] Kumar M, Sharma SC. PSO-based novel resource scheduling technique to improve QoS parameters in cloud computing. Real-world optimization problems and meta-heuristics. Neural Computing and Applications. 2019; **134**(16):1-24. DOI: 10.1145/3302505.3310069

[2] Zhang C, Kuppannagari SR, Kannan R, Xiong C, Prasanna VK. A cooperative multi-agent deep reinforcement learning framework for real-time residential load scheduling. Association for Computing Machinery. ACM. 2019. pp. 59-69. ISBN: 978-1-4503-6283-2/19/04 DOI: 10.1145/3302505.3310069

[3] Song Z, Liu B, Cheng H. Adaptive particle swarm optimization with population diversity control and its application in tandem blade optimization. Journals of Mechanical Engineering Science, Sage. 2018;**233**(6):1-17

[4] Dulhare U. Prediction system for heart disease using Naïve-Bayes and particle swarm optimization. Biomedical Research. 2018;**29**(12):2646-2649

[5] Shah J, Zhang C. Co-optimizating multi-agent placement with task assignment and scheduling. Proceedings of the Twenty-Fifth International Joint Conference on Artificial Intelligence. ACM. 2016. pp. 3308-3314. ISSN No: 978-1-57735-770-4

[6] Khalil KM, Abdel-Aziz M, Nazmy TT, Salem A-BM. MLIMAS: A framework for machine learning in interactive, multi-agent systems. Procedia Computer Science. 2015;**65**:827-835. DOI: 10.1016/j.procs.2015.09.035

[7] Riedmiller M, Sperschneider V, Brockmann W, Nuchter A. Multi-agent reinforcement learning approaches for distributed job-shop scheduling problems. 2009

[8] Kanaga EGM, Valarmathi ML. Multi-agent based patient scheduling using particle-swarm optimization. Procedia Engineering. 2012;**30**:386-393

[9] Zhan Z-H, Zhang J, Li Y, Chung HS-H. Adaptive particle swarm optimization. IEEE Transactions on Systems, Man, and Cybernetics. 2009;**39**(6):1362-1381

[10] Zeng J, Jie J, Hu J. Adaptive particle swarm optimization guided by acceleration information. International Conference on Computational Intelligence and Security. IEEE, 1-4244. 2006. pp. 351-355

[11] Kennedy J, Eberhart R. Particle swarm optimization. IEEE Xplore. 2002. pp. 1942-1948

[12] Brauer W, Weiß G. Multi-machine scheduling – a multi-agent learning approach

Chapter 4

On an Approach to Knowledge Management and the Development of the Knowledge-Based Multi-Agent System

Evgeniy Zaytsev and Elena Nurmatova

Abstract

The chapter discusses the architecture of the Knowledge-Based Multi-Agent System (KBMAS) and describes the software agent models. The purpose and functional organization of the system software agents used for planning and management of computing resources of the KBMAS are considered. An approach to the applied software agent's development that integrates knowledge-based reasoning mechanisms with neural network models is proposed. The structure of the problem-oriented Multi-Agent Solver, including groups of reactive and cognitive software agents used to solve complex ill-formalized problems, is considered. The interaction diagram of reactive agents and the states and transitions diagram of cognitive agent of the computing node are given. The control scheme is shown that includes methods for determining the availability of microservices used by agents, reliability assurances and coordinated operation of the system's computing nodes. The method of reinforcement learning, the system of rules (productions), and the queries to the knowledge base are described. Methods of distribution of software agents in the KBMAS computing nodes, as well as construction of an optimal logical structure of the Distributed Knowledge Base, which has minimal information connectivity and ensures effective operation of the system on multicomputers, are proposed.

Keywords: distributed system, multi-agent system, knowledge-base, intelligent software agents, Fuzzy system, reinforcement learning, data localization optimization

1. Introduction

The Knowledge-Based Multi-Agent System is a Distributed Artificial Intelligence System that uses intelligent applied and system software agents. Knowledge-based reasoning mechanisms and artificial neural networks are integrated into the model of applied software agents, which are designed to solve complex, ill-formalizable problems [1–4].

System software agents are used to effectively manage computational processes and provide application software agents with access to information-computing resources of the multicomputer.

Applied software agents function using Event-Driven Microservices (EDM) — independent, autonomous resources designed as separate interacting processes with lightweight interprocess communications. Microservices can be implemented as autonomous processes or as functions, they may or may not store state.

A feature of the microservice architecture is that microservices are loosely coupled. EDM communicates not through request-response API (Application Programming Interface), but through events defined in event streams, which are neutral with respect to one technology or another. This allows you to choose the most appropriate tooling to implement each job, allowing you to achieve the required level of performance.

In today's EDM architecture, information is exchanged by issuing and consuming events, which may be transferred through simple notifications as well as complex structures with state support. The events are not destroyed when consumed, as in conventional messaging systems, but remain available to other consumers, who can read them as needed.

Microservices consume events from input event streams, process information, generate their own outgoing events, provide data for to implement a request-response access scheme, exchange information with third-party APIs, or perform other necessary actions.

Unlike Service-Oriented Architecture (SOA), which typically uses web service standards like SOAP (Simple Object Access Protocol), microservice architecture uses simpler protocols. A microservice can be designed as a stand-alone service on a PAAS (Platform As A Service), or it can be a process of its own Operating System.

In traditional Operating System architectures, information-computing resources are hidden behind universal APIs that do not allow the KBMAS developer to implement problem-dependent optimization. Effective implementation of KBMAS is possible on the basis of an exokernel OS and event-driven intelligent system software modules.

High performance is achieved through the implementation of special mechanisms for managing information and computing resources, as well as the possibility of direct access to hardware [5, 6].

Using the services of an exokernel OS, the KBMAS designer is able to choose or implement his own System Libraries (LibOS). For example, specialized VMM (Virtual Memory Management) or IPC (InterProcess Communication) modules defined in LibOS can work much faster than general-purpose software modules doing a similar job in a monolithic or microkernel Operating System. System software agents can effectively manage information-computing resources using exokernel OS services, which is the basis for creating high-performance knowledge processing systems.

Currently, in distributed systems, hypervisors are usually used instead of exokernel architecture. Hypervisors of the first type do not use OS services, they directly control the hardware. In hypervisors of the second type, instead of an exokernel that provides untrusted servers with a low-level interface to access computing resources, the host OS runs. Guest OSs are used instead of user (unprivileged) mode servers. The advantage of building an exokernel virtualization system instead of using hypervisors is that, in this case, an extra layer of mapping is eliminated.

Unlike a hypervisor, which must support disk address translation tables (and other tables to convert virtual resources to physical resources), there is no need for such reassignment when using an exokernel OS. The exokernel only needs to keep track of which Virtual Machine (VM) has been given certain hardware resources. The exokernel architecture does not require creating a copy of a real computer and isolating virtual machines from physical resources. In this case, each VM is provided with a subset of the real computer resources. The exokernel OS operates in privileged mode, distributing computing resources between VMs and controlling them so that none of the machines tries to use not intended for this VM computing resources.

Virtual machines can serve as containers that contain software agents (processes and threads) and their surroundings. It is easier to provide mobility of software agents together with their VM than to move individual agents around. When using a virtual machine, the local group of software agents is moved together with the necessary environment for it (configuration files, system tables, etc.).

Development of KBMAS on the basis of exokernel OS includes creation of system and applied software agents, definition of their states and actions, as well as events (messages), delivery environment properties, and other characteristics describing the agents and their interaction. Development of a problem-oriented KBMAS can be performed using a special Multi-Agent-KB5 software toolkit [4], which allows knowledge engineers to design a distributed intelligent system based on high-level abstractions implemented by high-performance specialized system software modules. Using the Multi-Agent-KB5 toolkit, a knowledge engineer can create groups of applied software agents that act rationally, are able to respond in a timely manner to environmental events and learn in this environment.

In a broad sense, rationality is the ability to do the right thing. Ideal rationality, that is, choosing the optimal (best) action in a given situation, is not always achievable and may require large computational resources. The concept of rationality in the KBMAS uses to both applied and system software agents.

In logic programming paradigm, the rational behavior of an applied software agent is realized by means of logical inference methods (resolutions and unifications). A software agent can act rationally without using logical inference. In some situations, a reflex action may be more successful than a slower action taken after logical inference.

Problem-solving in the KBMAS is done by decomposing a complex problem into subtasks, which are jointly solved by rational applied software agents. Horizontal and vertical decomposition is used. Horizontal decomposition results in a multi-connected system with a flat structure. Vertical decomposition results in a hierarchical system with multiple levels. The levels are vertically subordinate to each other and have their own goals and functions, the implementation of which is aimed at achieving the global goal of the intelligent system.

Two types of applied software agents are used in the KBMAS to solve applied problems: cognitive and reactive. Mathematical models of these types of agents are described in [4].

KBMAS is an emergent system that implements the principle of self-organization. In the self-organizing KBMAS software agents are capable of making decisions under conditions of incompleteness, vagueness, and fuzziness of knowledge.

2. Structural and functional organization of the KBMAS

In traditional Knowledge-Based Systems, problems are solved by a single intelligent solver. This intelligent solver is designed as a monolithic application. It is assumed to use a complete and consistent Knowledge Base and has a global view of the problem. This model uses monotone logic (closed-world), the intelligent solver search by AND/OR-connection (reduction) graph (**Figure 1**).

The reduction graph shown in **Figure 1** corresponds to the following fragment of the formalized description of the problem domain (in the language of logical programming):

RCSF(A,B,C,...,G,F) :- RCSF(A,B,C,...,G, f1); ... ; RCSF(A,B,C,...,G, fn).
RCSF(A,B,C,...,G,f1) :- RCS(A,B,C,f1), VS(G,A,B,C,f1).
RCS(A,B,C,F) :- SOVA(A,B,F), SOVB(B,C,F), SC(C,A).
SOVA(X,Y,F) :- SA(X,Y), FR1(X,Y,F).
SOVB(K,L,F) :- SB(K,L), FC(L,F).
FR1(G,M,R) :- FA(G,R), FB(M,R).

There are effective parallel algorithms for reduction graph processing. However, these algorithms can be used only in Shared Memory Processors systems (SMP). These algorithms are not suitable for multicomputers, which use a completely different model of concurrency — message-based distributed computing.

To implement parallel search algorithms on the reduction graph in multicomputer (cluster) systems, as an option, it would be possible to organize a single virtual address space (Distributed Shared Memory, DSM) using page swap. However, in the case of the DSM mechanism, which is implemented by the Operating System or middleware, system performance is low. A more complex but predictable message-based model is preferable.

The Knowledge-Based Multi-Agent System uses a network of cooperating software agents instead of a single intelligent solver performing a parallel search on the

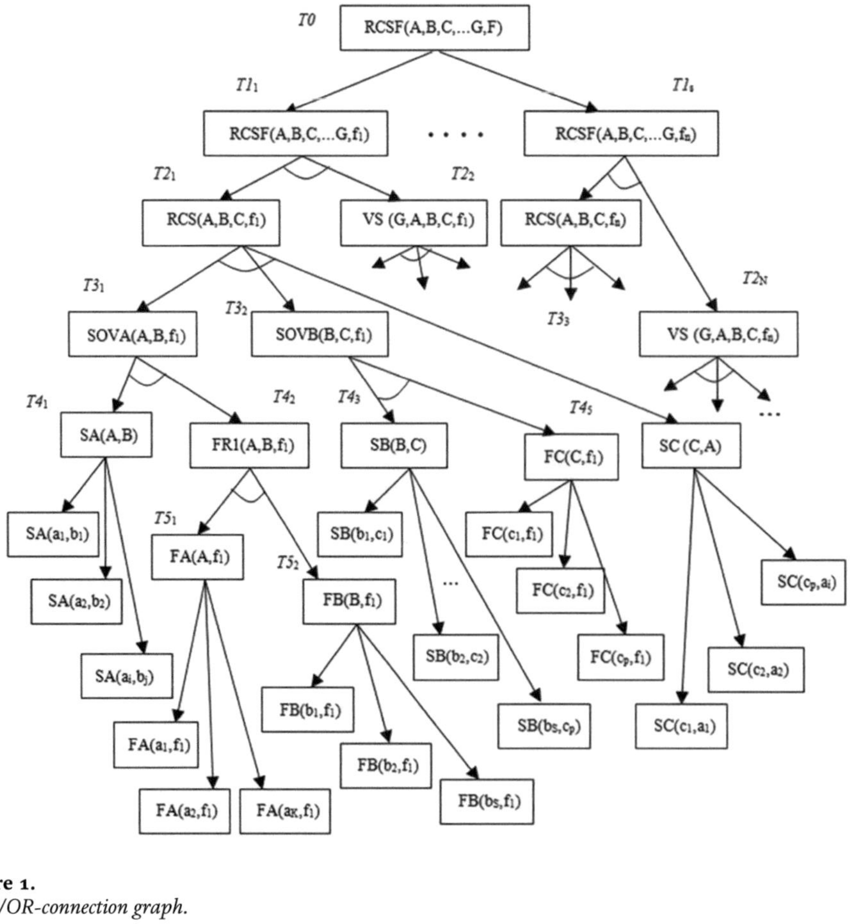

Figure 1.
AND/OR-connection graph.

reduction graph. Each individual software agent has only partial knowledge of the problem and can solve only some subtask.

The KBMAS integrates models, methods, and tools of Distributed Artificial Intelligence, parallel computing, and Event-Driven Microservices technology. The software agents of the KBMAS are loosely coupled intelligent software modules that can be distributed, often on a large scale (**Figure 2**).

The use of decoupled components is a basic requirement for successful scaling. The opportunity to realize decoupled software modules is provided by the methodology of Object-Oriented Analysis and Design. The idea of decoupling underlies most of the of Object-Oriented patterns, which can be successfully used to create a Knowledge-Based Multi-Agent System.

Computing nodes of the KBMAS are multiprocessors with shared memory. Software agents of the same node communicate with each other using a Local InterProcess Communication (LIPC). To improve performance the LIPC is implemented using specialized system libraries (LibOS). Software agents of the different nodes communicate through message exchange implemented using standard libraries and middleware.

Figure 3 shows the structure of a Multi-Agent Solver for one of the KBMAS nodes. Two types of applied software agents are used in the Multi-Agent Solver: cognitive and reactive. Applied software agents processing the domain knowledge use special Cognitive Data Structures (CDS). Four types of methods are implemented for work of applied software agents with the knowledge base: comparison (CMP), association (ASS), analysis (ANS), and specification (VAL). CMP method is called when

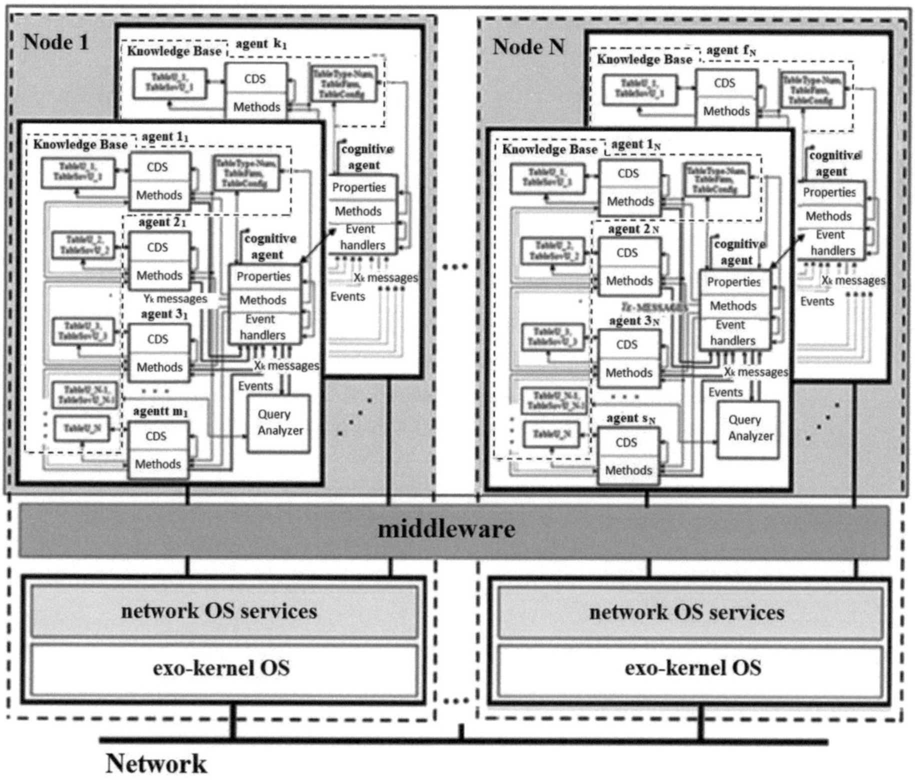

Figure 2.
Structure of the KBMAS.

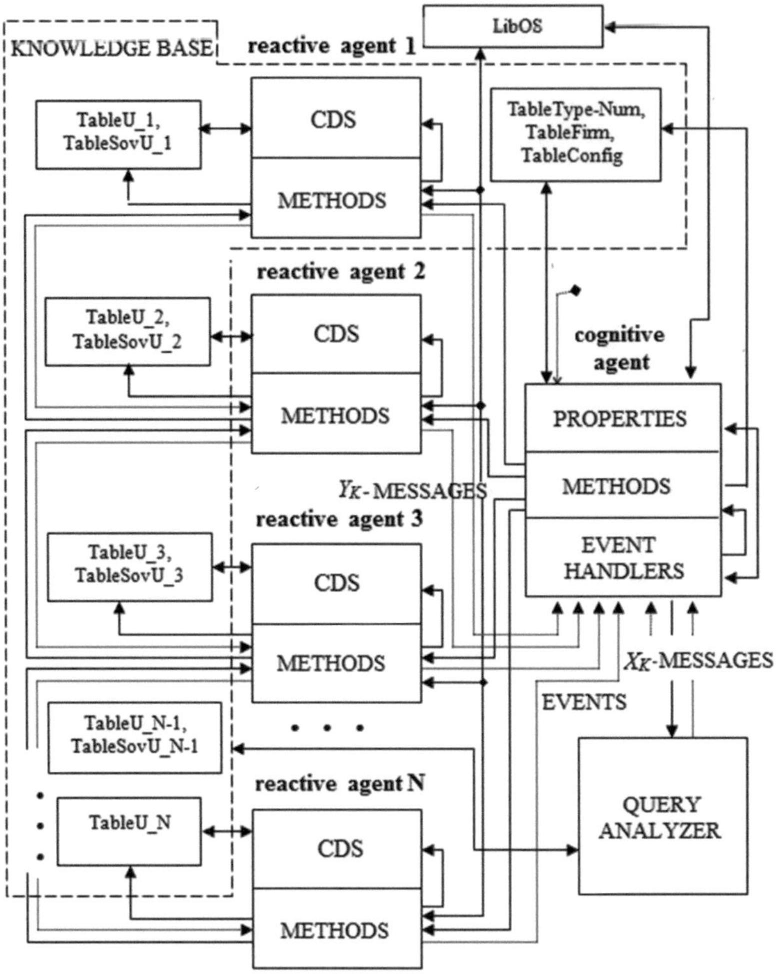

Figure 3.
Multi-Agent Solver of the KBMAS node.

comparing events or objects; ASS method is used to get answers to queries about relations between objects and events; ANS method implements logical analysis of events. For object specification (VAL-method) can be used both clear queries to the knowledge base and fuzzy queries [7].

Different types of membership functions can be used to implement fuzzy queries. **Figure 4** shows the examples of membership functions (μ_s) for two linguistic variables: Correctness and Completeness.

Figure 5 shows the result of a fuzzy query that uses these membership functions.

In the problem-oriented Multi-Agent Solver presented in **Figure 4**, the priorities of the applied software agents are set according to the sequence number of the software agent. In this case, the first software agent uses the tables TableU_1 and TableSovU_1. The software agent with number N has the lowest priority and is associated with table TableU_N.

Cognitive applied software agent coordinates the work of a group of reactive software agents. As an example, **Figure 6** shows the diagram of the states and transitions of one of the cognitive applied software agents.

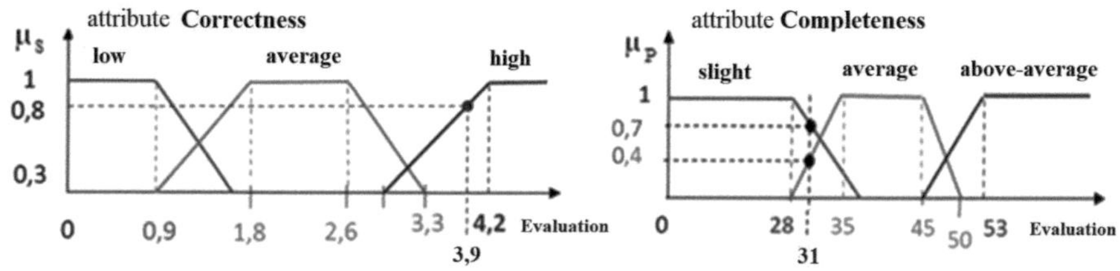

Figure 4.
Examples of the membership functions.

ID	Denomination	Completeness	Correctness
1	N1	47	3,2
2	N2	25	1,5
3	N3	26	3,9
4	N4	31	4,8
5	N5	23	2,5

SELECT *from TableU_1 where (Correctness = "high" AND Completeness = "slight")

ID	Denomination	Completeness	Correctness	FP
3	N3	26	3,9	0,8
4	N4	31	4,8	0,7

Figure 5.
Example of the result of a fuzzy query.

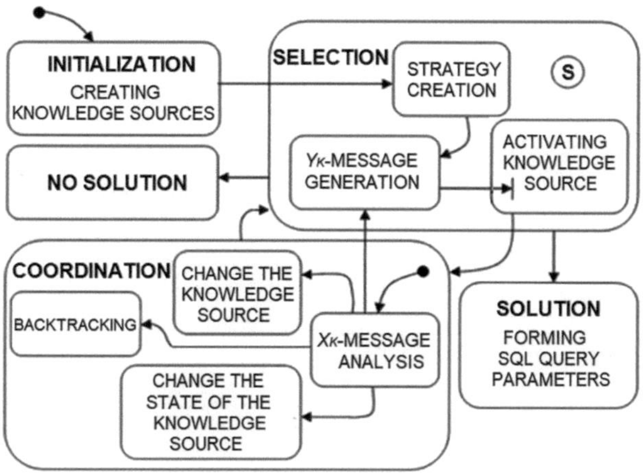

Figure 6.
State and Transition Diagram.

After initialization, the transition to the "Selection" state occurs, in which the cognitive software agent selects the necessary knowledge source, taking into account the informative signals from the reactive agents. The cognitive agent then goes into a

"Coordination" state in which it coordinates the actions of the reactive software agents. If at the next step, the reactive agents do not find a coordinated solution, cognitive agent backtracking to the previous state of partial solution to the problem.

The diagram of one possible option of interaction between the reactive agents of a node, each of which is connected to only one neighbor, is shown in **Figure 7**.

The organization of the Multi-Agent Solver is described in more detail in the paper [4].

Computing node of the KBMAS can use the concurrency model in which several computing threads are in the execution state, and only one of the threads is actually executing on the processor at any given time. This concurrency model uses shared memory to exchange information. Competitive threads are described by the consistency model, which defines the order in which operations performed by local agents in a node should be executed, and the order in which the results of these operations should be transmitted to the group members.

In addition to competitive concurrency computations, simultaneous concurrency computations can be implemented in the KBMAS nodes. To implement parallel computing, Uniform Memory Access (UMA) multiprocessors are usually used. In this case, the whole set of software agents is divided into subsets (groups). Agents that belong to different groups can act simultaneously.

The execution of computing processes (threads) associated with groups of applied software agents is coordinated by system software agents, which provide access to information-computing resources of the multicomputer.

The agents are dynamically divided into groups using the compatibility matrix S and the inclusion matrix R. The compatibility matrix S has the following form (1):

$$S = \begin{bmatrix} 0 & s_{12} & s_{13} & \cdots & s_{1M} \\ s_{21} & 0 & s_{23} & \cdots & s_{2M} \\ s_{31} & s_{32} & 0 & \cdots & s_{3M} \\ \cdots & \cdots & \cdots & \cdots & \cdots \\ s_{M1} & s_{M2} & s_{M3} & \cdots & 0 \end{bmatrix} \begin{matrix} S_1 \\ S_2 \\ S_3 \\ \\ S_M \end{matrix} \qquad (1)$$

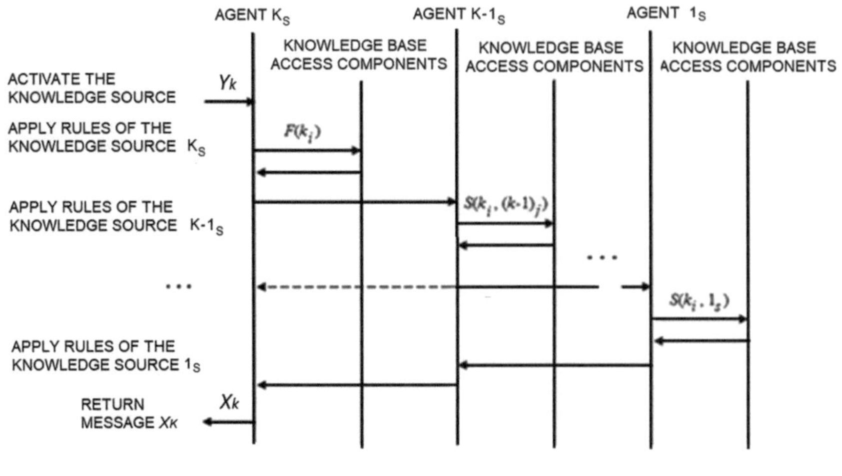

Figure 7.
Interaction scheme of the reactive software agents.

where s_{ij}=1, if the agents A_i and A_j use different computing resources and work in parallel, otherwise s_{ij}=0.

The distribution of agents into groups is based on the inclusion matrix R (2):

$$R = \begin{bmatrix} r_{11} & r_{12} & \cdots & r_{1M} \\ r_{21} & r_{22} & \cdots & r_{2M} \\ \cdots & \cdots & \cdots & \cdots \\ r_{H1} & r_{H2} & \cdots & r_{HM} \end{bmatrix} \begin{matrix} R_1 \\ R_2 \\ \\ R_H \end{matrix} \tag{2}$$

where M is the number of agents, H is the number of groups. r_{ij} = 1 if the agent A_i is included in the group Y_j. The agent A_i is included in the group Y_j if $S_i \cap R_j = \varnothing$, that is the matrix rows do not intersect. For optimal partitioning into subsets, it is necessary to consider the functional features of the software agents, their requirements for computing resources, as well as know the structural organization of the KBMAS node used to implement parallel computations.

In the process of solving subtasks, software agents use microservices, which can be duplicated on different nodes of the computing system. The system software agents distribute the computational load and control the microservers based on the data provided by the agents-monitors.

The behavior of applied software agents becomes more rational by repeatedly solving the same problem. The applied software agent learns through a series of rewards and punishments. The agent's actions change the environment to a new state. The environment returns the next state and reward to the agent. The cycle "state → action → reward" repeats until the problem is solved (**Figure 8**).

Signals s_t correspond to a state, a_t to an action, r_t to a reward at time t. The strategy according to which the agent chooses actions is a function that maps a set of states into a set of actions. The agent's task is to choose (by trial and error) the best action that maximizes the target function, which is the sum of the rewards received by the software agent.

Various reinforcement learning algorithms can be used. The most effective is the Actor-Critic algorithm [8–10] in which the strategy generates action, and the value function critiques the actions.

The basis of reinforcement learning is function approximation. As a method of function approximation, KBMAS uses a multilayer neural network that models both policy functions and value functions.

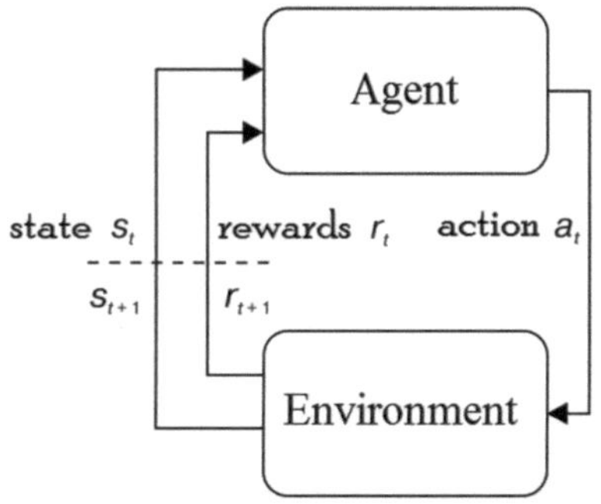

Figure 8.
The reinforcement learning control loop.

The advantage function A=Q(s,a)-V(s) is used to generate reinforcing signals. Benefit V(s) which can be obtained by achieving a particular state (s) is evaluated by the agent before the action, and the value function of the action Q(s,a) — after the action has been taken.

3. Aggregation of distributive information structure

Created information data structures can have a large volume and dimension, so their loading and implementation is carried out in fragments. Logically interrelated data should be divided into a number of clusters that have the smallest interconnection under constraints on the dimensionality of clusters, as well as on the degree of semantic proximity of logical records included in the clusters. No less important is the question of choosing the type of data storage systems used [11, 12].

Let us introduce the variable B^i_{kj}, which characterizes the handling by the k-th query of the i-th information element, which is in the j-th logical record. Variable $X_{ij} = 1$ if the i-th data partition is selected in the j-th logical record; $X_{ij} = 0$ otherwise. Variable $a_{ik} = 1$ if the i-th data partition is included in the k-th query; otherwise $a_{ik} = 0$.

Variable B^i_{kj} characterizes the use of the j-th logical record by the k-th query (3):

$$B^i_{kj} = \begin{cases} 1,\ when \sum\limits_{i=1}^{I} a_{ik}x_{ij} \geq 1; \\ 0,\ when \sum\limits_{i=1}^{I} a_{ik}x_{ij} = 0; \end{cases} \tag{3}$$

Note that the results obtained in solving such a problem are important from a practical point of view, given the constraints for designing an acceptable data structure and the ability to generate fast queries for sampling and editing distributional structures.

Let's analyze this algorithm step by step.

4. Approximate algorithm for distributing data clusters between the server and local network clients

At this stage, the distribution of data batches for storage and processing is determined by the criterion of minimum total traffic.

We reduce the canonical graph of the data structure to an uncoupled graph and calculate the weight of each data batch, summarily consisting of the weight of the batch itself and the weight of arcs (links), taking into account the requirements of the network clients (4):

$$W_i = W_i^{part} + W_{iq}^{link} \tag{4}$$

where, W_i^{part} — total weight of data partitions (5); W_{iq}^{link}— the weight of the arcs of the canonical data structure graph (6).

$$W_i^{part} = \sum_{k=1}^{k_0} \sum_{p=1}^{p_0} \gamma_{kp}^{Q} \delta_{kp}^{Q} \vartheta_{pi} \tag{5}$$

$$W_{iq}^{link} = \sum_{k=1}^{k_0} \sum_{p=1}^{p_0} \gamma_{kp}^{Q} \delta_{kp}^{Q} \vartheta_{pi} \sum_{q \neq i}^{I} \vartheta_{pq} a_{iq}^{G} \tag{6}$$

The weight of the i-th data batch is calculated by the formula (7):

$$W_i = \sum_{k=1}^{k_0} \sum_{p=1}^{p_0} \gamma_{kp}^{Q} \delta_{kp}^{Q} \vartheta_{pi} \left(1 + \sum_{q \neq i}^{I} \vartheta_{pq} a_{iq}^{G} \right) \tag{7}$$

where, γ_{kp}^{Q}—the frequency of generation of user requests; δ_{kp}^{Q}— elements of the matrix for generating user queries; ϑ_{pi}— the matrix for using data partitions when executing queries; a_{iq}^{G}— the semantic contiguity matrix of data partitions.

In the next step, the local network graph is converted into an unconnected graph with the calculation of the weight of each node (8):

$$W_r = t_r + \sum_{m \neq r}^{R_0} t_{rm} \tag{8}$$

where t_r — the total average duration of data processing in the r-th node, consisting of the time of decomposition of the query into subqueries, route selection and connection establishment, etc.;

t_{rm} — the average duration of data transmission between nodes, determined based on the matrix of logical distances between the servers of the nodes of the local network.

Next, the matrix $V = \|\omega_{ir}\|$ is formed, the elements of which are the Cartesian product of the weights of each node by the weights of each data partition (9):

$$\omega_{ir} = W_i \times W_r \ \text{ for } i = \overline{1,I}; r = \overline{1,R_0}. \tag{9}$$

In the final step of the first stage, the problem (10)

$$\min_{\{x_{ir}\}} \sum_{i=1}^{I} \sum_{r=1}^{R_0} \omega_{ir} x_{ir} \tag{10}$$

is solved under the constraints:

- by the number of data partitions, the localization of which is possible on one node (11)

$$\sum_{i=1}^{I} x_{ir} \leq N_r, r = \overline{1,r_0} \tag{11}$$

- on the permissible redundancy of groups by network nodes (12)

$$\sum_{r=1}^{r_0} x_{ir} \leq M_i, i = \overline{1,I} \tag{12}$$

- on the amount of available external memory of the data storage system (13)

$$\sum_{i=1}^{I} x_{ir}\rho_i\pi_i \leq \eta_r^{ROM} \tag{13}$$

where, ρ_i — the vector of group lengths in bytes; π_i — the vector of number of instances in groups; η_r^{ROM} — the amount of available memory on the server of the r-th host; $x_{ir} = 1$, if the i-th data partition is included in the r-th network node; $x_{ir} = 0$ – otherwise.

5. Problem of distributing data parties of each agent to the types of logical records

The problem posed at this stage is solved taking into account the criterion of the least total time of local data processing in each network node by agents. The number of aggregation tasks for this stage is determined by the number of local network nodes.

The initial data are the subgraphs of the graph of the canonical data structure, as well as the temporal and volumetric characteristics of the subgraphs of their canonical structure, a set of requests from users and network nodes.

The aggregation problem is solved here using approximate algorithms with restrictions:

- on the number of groups (14):

$$\sum_{i=1}^{I} x_{it} \leq F_t,\ \forall t = \overline{1,t_0},\ where,\ F_t - number\ of\ groups\ in\ t - record, \tag{14}$$

- on the non-repeatability of including groups in a record (15):

$$\sum_{t=1}^{t_0} x_{it} = 1,\ \forall i = \overline{1,I} \tag{15}$$

- on the cost of information storage.

The main cost characteristics of the distributive data structure are the cost of storing E_{ds} information; the cost of executing E_{run}^Q requests and transactions at a given time interval; the cost of transmitting information via E_{run}^C communication channels.

The sum of these components determines the total cost (16):

$$E = E_{ds} + E_{run}^Q + E_{run}^C \tag{16}$$

The cost of storing distributed information is determined by the physical volume of information Vcr and the cost of storing a unit of information volume (one logical record) on the server. If we assume that the cost of storage in all nodes of the local network is a constant value, then $E_{ds} = V_L \bullet k_{ds}$.

That is, the product of the logical volume of stored information (V_L) and the coefficient that takes into account the storage capacity on the media when organizing the database (k_{ds}; in practice, it is approximately equal to 1.2–1.5).

The cost of executing multiple user requests at a given time interval is the sum of the cost of servicing multiple user requests on servers and the cost of transmitting information through communication channels during the execution of user requests.

The cost of performing transactions at a given time interval is also determined by the sum of the cost of performing steps (tasks) of transactions on server nodes and the cost of transferring transaction requirements to server nodes, fixing transactions and removing locks.

- from the total time of servicing operational requests on servers (17):

$$\sum_{rc=1}^{r_0} \sum_{t=1}^{t_0} B_{pr}^{t} \bullet (t_q + t_l) < T_p \tag{17}$$

where T_p — additional service time of the p-th operational request; t_q — average duration of generation of one request (or transaction task step); t_l — average processing time for one logical record on a local network host/server.

Variables B_{pr}^{t} determine the types of logical records used by the p-th query in the r-th node of the computing system.

As a result, logical database structures are determined for each network node.

6. Localization of data by network nodes

This step uses the results of the previous steps and the characteristics of the data warehouse.

As a result of the proposed algorithm, localization matrices for a set of batches of data are formed by types of logical records (the result of the first stage), and then groups of records by local network nodes. In this case, the running time of the algorithms is additionally estimated.

7. Conclusion

The chapter examined the structural and functional organization of the Multi-Agent System, which uses intelligent applied and system software agents. To support the process of developing a problem-oriented KBMAS based on the considered agent-based models, the Multi-Agent-KB5 toolkit is used. This toolkit includes interactive wizards and property panels that allow creating groups of applied (reactive and cognitive) software agents. Inclusion of system software agents into KBMAS, which implement algorithms for planning and management of computational resources, taking into account the specifics of the interaction of applied agents, increases the performance of the system. KBMAS performance is also improved by optimizing the logical structures of the distributed knowledge base.

Author details

Evgeniy Zaytsev* and Elena Nurmatova
MIREA—Russian Technological University, Moscow, Russia

*Address all correspondence to: zajcev@mirea.ru

References

[1] Wooldridge M. An Introduction to Multi-Agent Systems. 2nd ed. John Willey & Sons Ltd; 2009. p. 488. ISBN: 978-0-470-51946-2

[2] Baranauskas R, Janaviciute A, Jasinevicius R, Jukavicius V. On multi-agent systems intellectics. Information Technology and Control. 2015;**1**:112-121

[3] Houhamdi Z, Athamena B, Abuzaineddin R, Muhairat M. A multi-agent system for course timetable generation. TEM Journal. 2019;**8**:211-221

[4] Zaytsev EI, Khalabiya RF, Stepanova IV, Bunina LV. Multi-agent system of knowledge representation and processing. In: Proceedings of the Fourth International Scientific Conference "Intelligent Information Technologies for Industry" (IITI'19). Springer; 2020. pp. 131-141

[5] Darweesh S, Shehata H. Performance evaluation of a multi-agent system using Fuzzy Model. In: 1st International Workshop on Deep and Representation Learning (IWDRL). 2018. pp. 7-12

[6] Aly S, Badoor H. Performance evaluation of a multi-agent system using Fuzzy Model. In: 1st International Workshop on Deep and Representation Learning (IWDRL). Cairo; 2018. pp. 175-189

[7] Zaytsev EI. Method of date representation and processing in the distributed intelligence information systems. Automation Modern Technologies. 2008;1:29–34

[8] Red'ko VG. Evolyutsiya, neyronnyye seti, intellekt: Modeli i kontseptsii evolyutsionnoy kibernetiki. -M.: "LIBROKOM". 2013

[9] Graesser L, Keng WL. Foundations of Deep Reinforcement Learning. Addison-Wesley Professional; 2020. p. 416. ISBN: 978-0135172384

[10] Raj JS, Ananthi JV. Recurrent neural networks and nonlinear prediction in support vector machines. Journal of Soft Computing Paradigm (JSCP). 2019;**1**: 33-40

[11] Batouma N, Sourrouille J. Dynamic adaption of resource aware distributed applications. International Journal of Grid and Distributed Computing. 2011; **4**(2):25-42

[12] Nurmatova EV, Gusev VV, Kotliar VV. Analysis of the features of the optimal logical structure of distributed databases. In: Collection of works the 8th International Conference "Distributed Computing and Grid-technologies in Science and Education". Dubna; 2018. p. 167

Section 3

Advanced Developments in Multi-Agent Technologies and Machine Learning Creating Potential for Their Further Integration

Chapter 5

Modeling Electric Vehicle Charging Station Behavior Using Multiagent System

Jaslin Shaleem Khan,
Malligama Arachchige Uditha Sudheera Navaratne
and Janaka Bandara Ekanayake

Abstract

Agent-based models(ABMs) are a type of simulation in which a large number of self-sufficient agents interact in a way that combines stochastic and deterministic behavior. Recently, there have been reestablished interests in utilizing multiagent systems (MASs) to get more granular data relating to specific conditions. MESA is an ABM framework for Python. It enables users to quickly develop ABMs with built-in core components, view them with a browser-based interface, and evaluate their findings with Python's data analysis capabilities. This chapter depicts an ABM of a photovoltaic (PV)-powered electric vehicle (EV) charging station in a university car park modeled using MESA. The goal is to determine the preliminary requirements for PV-powered EV charging stations, which would result in increased PV and cost benefits.

Keywords: agent-based model (ABM), MESA, multiagent system (MAS), photovoltaic (PV), EV charging station

1. Introduction

Agent-based applications are becoming the mainstream in a wide range of domains, including e-commerce, logistics, supply chain management, telecommunications, healthcare, engineering, and manufacturing, as technology advances [1]. Multiagent systems (MASs) emerge as new software technologies that combine a number of artificial intelligence (AI) techniques. It provides a more efficient and natural alternative to building intelligent systems, thereby providing a solution to the current complex real-world problems that must be solved. Autonomy, complexity, adaptability, concurrency, communication, distribution, mobility, security and privacy, and openness are some of the properties of MASs [2, 3]. The concept of MAS is also a trending technology in power engineering applications, such as power system restoration, power system optimization, market simulation/electricity trading, and smart grid control [4]. The MAS's ability to deal with complex problems through agents, which has been highlighted in various research works, is the basis of using the MAS in many applications.

MAS-based models are used globally to implement demand-side management (DSM) systems, cost optimization, robustness management in microgrids (MGs), and controlling voltage and thermal constraints in distribution networks.

In [5], an MAS-based online voltage monitoring system was proposed. In [4, 6], a dual-layered advanced control and security system based on the MAS, as well as an automated meter reading facility in a smart grid distribution network, was proposed. DSM has been activated from a global research standpoint by maximizing the use of distributed energy sources, environmentally friendly technologies, optimizing algorithms, and the implementation of renewable energy (RE) resources [7–10]. The use of renewable energy sources in smart grid distribution systems, as well as the multi-MG model, lowers consumer electricity costs [11, 12].

As the number of electric vehicles (EVs) increases, better charging infrastructure is required to provide the necessary energy for mobility with the cost benefits. MASs have seen a surge in popularity in recent years, and they are now widely used in EV-based power system research studies. In [13], a multiagent system (MAS)-based modeling tool has been proposed to assess the effects of EV charging on Singapore's energy grid. This study looked into the effects of EV temperature (air conditioning) and EV charging load during the charging. In [14], a state of charge (SOC)-based charging algorithm has been suggested, which is divided into two categories: controlled and uncontrolled charging (vehicle to grid—V2G) and grid to vehicle (G2V). This reduces the number of cars that run out of power on their next trip. In [15], a decentralized and intelligent MAS for controlling and managing EV charging in low-voltage (LV) distribution networks was presented. In this context, three case studies were investigated: without EV charging regulation-uncoordinated case or dumb charging; with EV charging regulation and without the voltage charging control; and with EV charging and voltage droop control. The simulation results showed that charging regulation provides significant benefits in terms of voltage control when compared to the other two situations. The proposed solution in [14] has been improved with active demand (AD) program management in [16]. It enabled the incorporation of EVs into the system while mitigating their negative impact on voltage regulation.

Lee et al. [17] discussed the impact of EVs on the electric grid; it was tested using real data from the "My Electric Avenue" initiative with ABM. It looked at how consumers' use of time-of-use (ToU) tariffs and vehicle range (battery capacity) preferences affect total and peak demand fluctuations at the local substation. In order to depict the complexity of an electric transportation system, an EV implementation based on agents was created [18]. After implementing various charging powers and charging patterns, the results were generated in a qualitative manner. In [19], the effects of influencing variables on EV charging demand, such as driver behavior, charging station location, and electricity pricing, were investigated. An ABM on Net Logo was used in this study to precisely simulate human aggregate behavior and its impact on load demand due to EV charging. In [20], the results of a study that used ABM simulation to simulate alternative charging infrastructure rollout techniques to allow for large-scale EV adoption were presented. The simulation included a variety of user types (residents, visitors, taxis, and sharing), as well as various types of charging infrastructure (level 2, clustered level 2, and fast charging).

An EV powered by an intermittent power source is an excellent way to ensure low-cost, emission-free mobility. An energy storage management hybrid optimization algorithm was presented in [21]. This algorithm flipped between deterministic and rule-based modes of operation depending on the power pricing band allocation. The cost degradation model and the levelized cost of photovoltaic (PV) power were

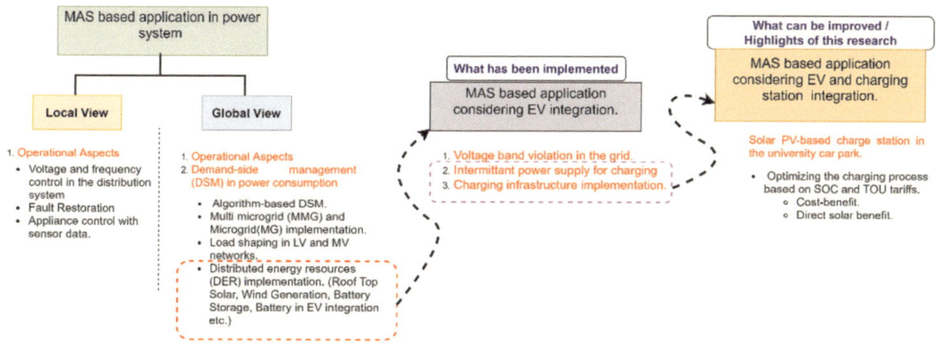

Figure 1.
An overview of this study and its state of the art.

combined in the case of PV-integrated charging stations with on-site energy storage systems. An agent-based charge station model utilizing renewable energy (RE) was proposed in [22]. Charging patterns were determined by scenarios including various RE capacities, policy interventions, limited versus unlimited charging capacity, social charging, and the existence or absence of central control. It was determined that in order to improve sustainable charging, policymakers should employ various incentives for different categories of EV drivers.

To aid comprehension of the state-of-the-art review and to clarify the study's contributions, **Figure 1** is added.

The rest of this chapter is structured as follows. Section 2 presents the modeling of an ABM of a PV-based charging station. Section 3 includes the simulation results and analysis of various scenarios. Section 4 draws the conclusion.

1.1 Solar PV-based EV charging station: Application of MAS

This chapter discusses the development of an agent-based EV charging station on university premises. It is primarily being developed as a solar PV-powered charging system. The objectives of this AB charging system are to optimize the benefits of cost and direct solar supply in a quantitative manner. This study makes two major contributions:

- The MAS simulation model for solar PV charging station for EV is proposed and developed using the motor vehicle database, which is assumed to be as EVs on university premises for the future energy distribution.

- The developed MAS provides a unique ABM simulation platform based on MESA software. It is able to incorporate energy optimization algorithms between interacting agents as SOC-based and TOU-tariff-based scenarios.

A survey was used to collect information about the available motor vehicles in the university car parks. A future EV integration database has been created in accordance with that.

1.1.1 Modeling of the system

The simulation model is built with a variety of agents, including an EV agent, a weather agent, a solar panel agent, a main control agent (MCA), a utility agent, a

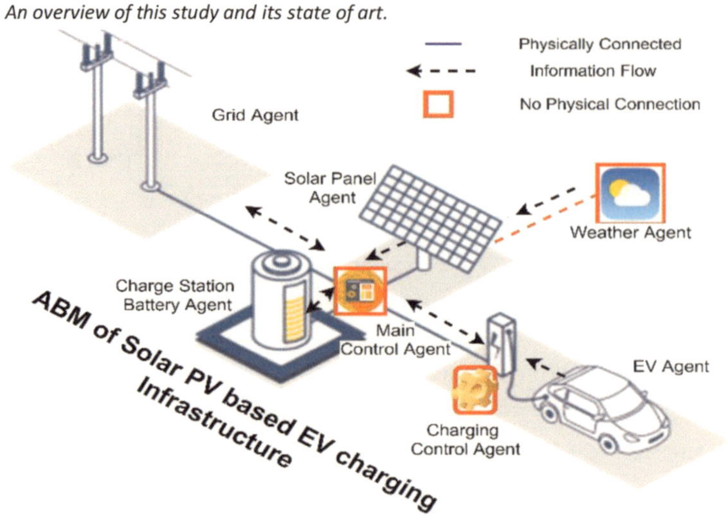

Figure 2.
The system architecture of the multiagent simulation platform.

charging control agent (CCA), and a charge station battery agent. The solar panel agent generates energy based on the weather condition's temperature and irradiance value, which can be accessed via the weather agent. The university weather broadcasting inverter portal was used to create the weather agent database. Each charging control agent in the developed agent-based system can manage the EVs' charge while taking into account energy prices and the EV agent's requirements (EVs' SOC during arrival, charging option, and arrival time), which is based on one of the following two scenarios: SOC-based or TOU-based tariffs. The EV agent represents the EV owner, who has the option of interacting/charging with the user interface in a range of ways. The charging control agent's energy scheduling is sent to the main control agent, which evaluates the overall energy supply agents' (utility agent, solar panel agent, and charge station battery agent) performance using optimizing algorithms. An illustration of the agent-based EV charging station system is presented in **Figure 2**.

1.1.2 Agents' validation

Each of the agents was validated with proper testing results before starting to simulate the system.

I. Solar panel agent: A PV array power calculation system of Homer Pro 3.14 was used to validate the solar panel agent. The results revealed the least amount of variation, which is included in **Table 1**.

The total PV array output Power (one day)	Simulation Model	Homer Model	Error%
	108.60 kWh	108.47 kWh	−0.123%

Table 1.
Total PV array output from the simulation model and Homer software.

	NEDC testing				WLTP testing			
	Simulation energy consumption (kWh/100 km)	**Standard energy consumption (kWh/100 km)**	**Error**	**Total distance (km)**	**Simulation energy consumption (kWh/100 km)**	**Standard energy consumption (kWh/100 km)**	**Total distance (km)**	**Error**
Without auxiliary	13.63	14.5	−6%	11.022	17.75	19.4	23.26	−8.46%
With auxiliary	14.52	14.5	0.138%		18.4	19.4		−5.15%

Table 2.
The NEDC and WLTP testing results of the ABM simulation.

		City_FC	HW_FC	Combined_FC
Standard energy consumption (kWh/100 km)		17.2	21.2	19.1
Standard fuel economy (km/Wh)		0.006424	0.005212	0.005879
Without auxiliary (simulation)	Fuel economy (km/ Wh)	0.0056755	0.005286	0.005500
	Error $\left(\frac{\text{STD}-\text{Simulation}}{\text{STD}}\right)\%$	−15.348	−0.778	−9.542
With auxiliary (simulation)	Fuel economy (km/ Wh)	0.00543755	0.005172	0.005318
	Error $\left(\frac{\text{STD}-\text{Simulation}}{\text{STD}}\right)\%$	−11.652	1.422	−6.439

Table 3.
The EPA testing results of the simulation.

Drive cycle	RMS current (A)	Test/simulation (SIM)	Error $\left(\frac{\text{Test}-\text{Simulation}}{\text{STD}}\right)\%$
US06	71.12	Test	1.38
	72.1	SIM	
HW	36.18	Test	0.06
	36.2	SIM	
USSD/FTP	23.85	Test	1.68
	24.25	SIM	

Table 4.
The EPA testing results for EV battery simulation.

II. EV agent: The EV agent was validated using the New European Driving Cycle (NEDC), Environmental Protection Agency (EPA), and Worldwide harmonized Light Vehicle Test Procedure (WLTP) tests. That was handled in two scenarios: no auxiliary device is used and certain auxiliary devices are turned on (with a power rating of 300 W based on the literature). The Nissan Leaf 2018 model was chosen to simulate testing with a test mass as the curb weight of 1573 kg and an additional payload of 100 kg. The results of the tests within the acceptable range are given in **Tables 2** and **3**.

III. Charge station battery agent: Lee et al. [23] explored the design and validation of a hardware-in-the-loop lithium-ion battery pack for EPA testing (US06, HW, and USSD/FTP). A two-time constant equivalent circuit battery cell model, as well as a lumped capacitance thermal model, was included in the battery pack model. To test the battery model in an EV as well as the charge station battery, a concurrently operating Li-ion battery from a Nisan Leaf 2012 vehicle power profile has been used. The results are given in **Table 4**.

2. ABM with MESA

2.1 MESA simulation software

The ABM of the EV charging station is modeled based on a multigrid scenario in MESA. MESA, a platform that provides a Python environment for agent behavior, is used to implement the agent-based control system. It contains the model (model, agent, schedule, and space), analysis (data collector and batch runner), and visualization (visualization server and visualization browser page) [24]. The portion of the Python-based MESA simulation source code is depicted in **Figure 3**.

The interaction of agents begins with the EV agent, which sends SOC information and charging requirements to its charging control agent (CCA). The request is then sent to the main control agent (MCA) by each CCA. The MCA sends requests to each energy resource, including the solar agent, charge station battery agent, and utility grid agent. The MCA then calculates the optimal charging schedule based on the charging scenarios (SOC-based and TOU-tariff-based). The MCA coordinates each step with each energy resource until the charging process is complete. Each iteration does its internal operations, such as calculations for energy management in MESA. Each agent is responsible for its own tasks.

Solar agent: It generates solar energy as per temperature and irradiance data, which is accessed from the weather agent.

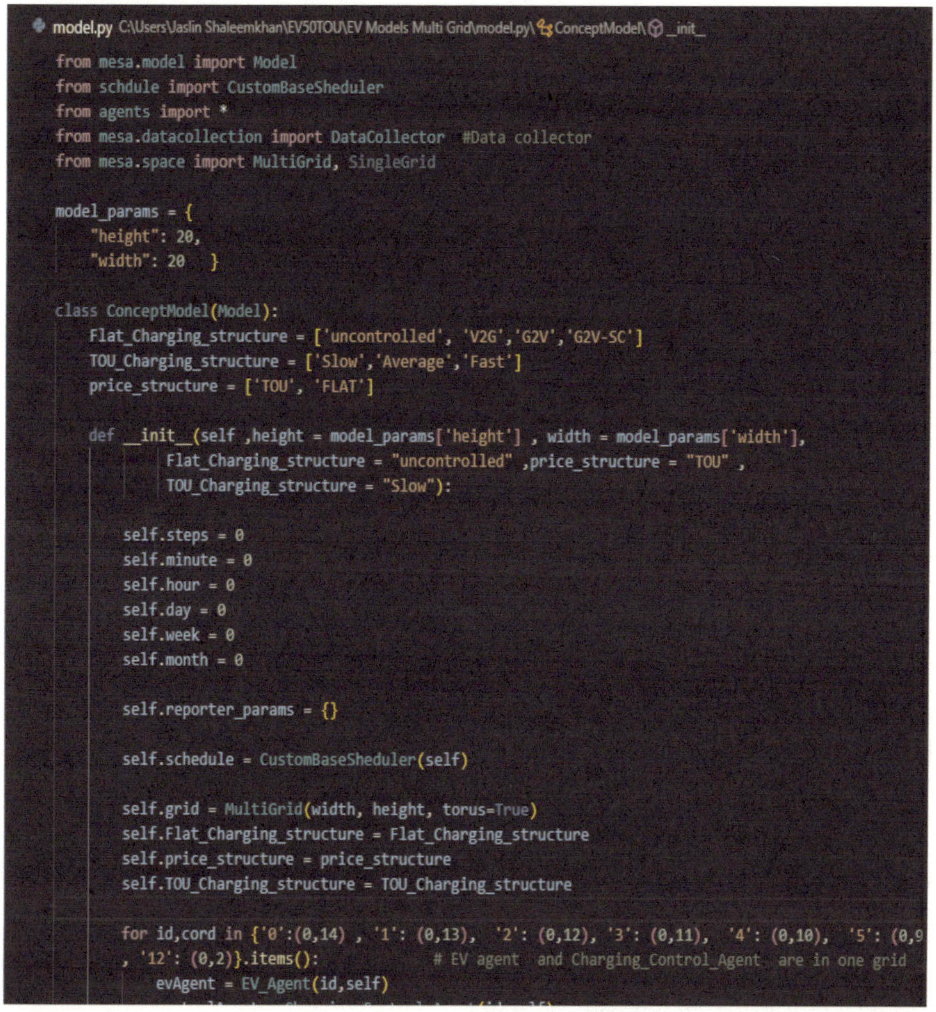

```
model.py C:\Users\Jaslin Shaleemkhan\EV50TOU\EV Models Multi Grid\model.py\ConceptModel\ __init__

from mesa.model import Model
from schdule import CustomBaseSheduler
from agents import *
from mesa.datacollection import DataCollector  #Data collector
from mesa.space import MultiGrid, SingleGrid

model_params = {
    "height": 20,
    "width": 20   }

class ConceptModel(Model):
    Flat_Charging_structure = ['uncontrolled', 'V2G','G2V','G2V-SC']
    TOU_Charging_structure = ['Slow','Average','Fast']
    price_structure = ['TOU', 'FLAT']

    def __init__(self ,height = model_params['height'] , width = model_params['width'],
                 Flat_Charging_structure = "uncontrolled" ,price_structure = "TOU" ,
                 TOU_Charging_structure = "Slow"):

        self.steps = 0
        self.minute = 0
        self.hour = 0
        self.day = 0
        self.week = 0
        self.month = 0

        self.reporter_params = {}

        self.schedule = CustomBaseSheduler(self)

        self.grid = MultiGrid(width, height, torus=True)
        self.Flat_Charging_structure = Flat_Charging_structure
        self.price_structure = price_structure
        self.TOU_Charging_structure = TOU_Charging_structure

        for id,cord in {'0':(0,14) , '1': (0,13),  '2': (0,12), '3': (0,11),  '4': (0,10),  '5': (0,9
        , '12': (0,2)}.items():            # EV agent  and Charging_Control_Agent  are in one grid
            evAgent = EV_Agent(id,self)
```

Figure 3.
The portion of source code—model.py.

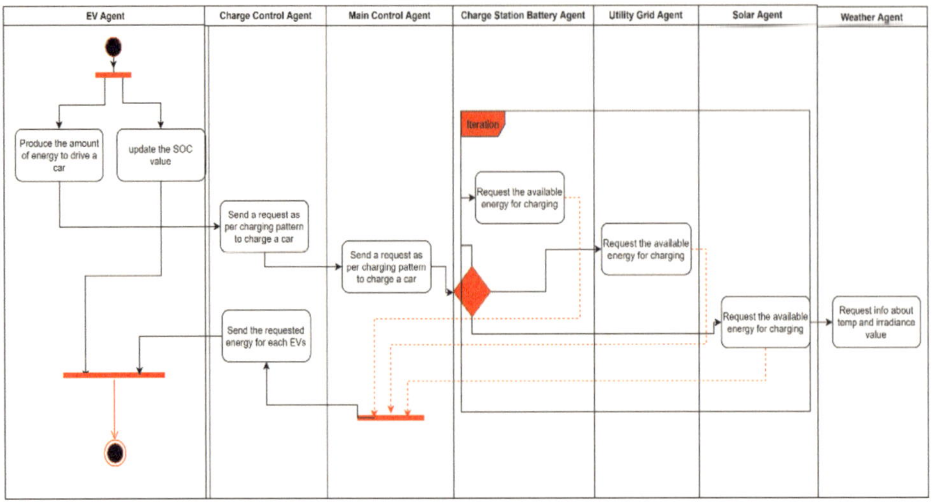

Figure 4.
The activity diagram of the ABM.

Utility agent: In this ABM, there is no power limit for the utility agent. It serves as a backup supply, allowing PV sources to sell excess energy and EVs to obtain power, depending on the power management strategy.

Main control agent: It serves as a coordinator, receives requests for EV charging from each charge pole agent, and performs internal calculations in accordance with management scenarios.

The ABM's activity diagram is depicted in **Figure 4**. It shows the information flows as well as agent internal operations.

2.2 Knowledge representation of agents

ABM with AI provides a real-time application that is extremely beneficial to all industries. Its popularity stems from its adaptability in different subfields (reasoning, knowledge representation, machine learning (ML), planning, coordination, communication, and so on). First, consider a few advantages of this combined technique: ABM drives emergent phenomena, precisely defines a natural system, and is flexible [2, 25].

Most complex real-world systems are only partially decomposable, and one solution would be to give the components the ability to decide on the nature and scope of their interactions at run time. Still, when combined with ML, ABM has the potential to create a new type of computing based on agents—by learning agents' behavioral patterns.

Many ABMs can easily incorporate various ML techniques such as genetic algorithms (GAs), neural networks (NNs), and Bayesian classifiers. It has two interlocked cycles for examining input, making decisions, and producing output. The ML algorithm uses the ABM as an environment for this framework, while the ABM uses the ML algorithm to maintain the agents' internal models [26]. The framework has described how ML techniques are used in ABM in **Figure 5**.

The weather agent in our solar-powered EV charging station model is updated with ML techniques to update its temperature and irradiance value. Excel is also used to

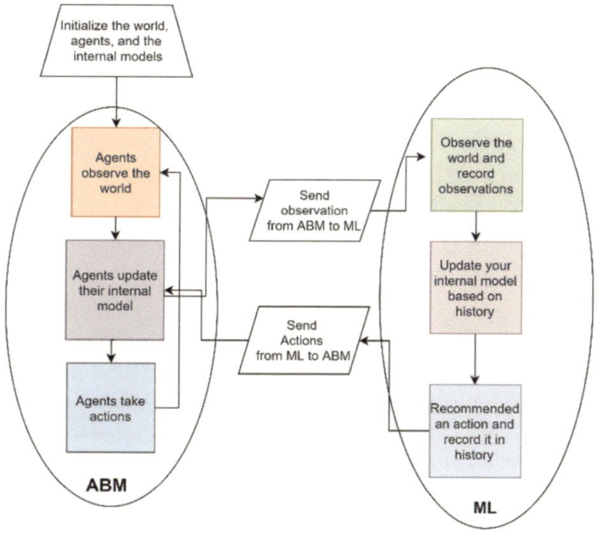

Figure 5.
The integrated cycles of ABM and ML [26].

generate the observed dataset from the faculty weather portal for an ML tool. Excel can be a valuable addition to your ML toolkit. This can aid in the visualization and analysis of smaller datasets.

2.3 Simulation and results

The simulations are evaluated in terms of direct solar benefit (DSB) and cost-benefit analysis. DSB defines the solar PV contribution to requested EV demand, which is calculated from Eq. (1).

$$\text{Total DSB}(\%) = \frac{\text{Direct solar supply for EV charging}}{\text{Total EV demand}} \times 100\% \qquad (1)$$

where

$$\text{Direct solar supply for EV charging} = \text{Available solar generation} - \text{Remaining solar energy}.$$

The system is simulated in 5-min intervals. This is simulated in two ways: SOC-based charging with a flat tariff and TOU-tariff-based charging.

The ABM of EV charging stations initially investigated using an SOC-based flat charging scenario. For EVs, three charging algorithms have been developed: uncontrolled, vehicle to grid (V2G), and grid to vehicle (G2V). TOU-tariff-based charging is simulated with slow, average, and fast charging options. The simulations have yielded numerical results for DSB and cost benefits.

2.3.1 SOC-based charging with flat tariff

Uncontrolled charging: This implies that when the EV is connected to the university charging station, the battery is charged until it reaches maximum SOC or disconnection.

Vehicle to grid (V2G): It converts EVs into energy storage systems, allowing any excess energy stored in the EV's battery to be injected back into the grid, which is enabled by the SOC values of EV. When the EV's battery pack SOC falls below 50%, it begins charging by fast charging (30 kW); otherwise, it charges by average charging (6.6 kW) until the EV reaches 80%. If the EV's SOC is greater than 80%, the energy in the EV battery can be pushed back into the electrical grid.

Grid to vehicle (G2V): It allows the EVs to charge with controlled charging while parked in the university car park. If the EV's SOC falls below 50%, it immediately begins charging either through fast charging or through average charging.

Figures 6–8 show the simulation of the above three charging scenarios. GE—grid energy, BSE—battery storage energy, and SE—solar energy.

When the PV energy supply is insufficient to fully charge the EVs, the stationary storage charges the EV, and the energy is supplied by the public grid. The main drawback is that PV energy supply does not completely benefit EV charging and that reliance on the public grid increases when charging is uncontrolled.

This ABM is modeled to serve as a university charging station (workplace charging). A large number of EVs are likely to be parked for a longer duration. To increase the direct solar benefit of EV charging, slow charging (2.3 kW) is combined with G2V charging.

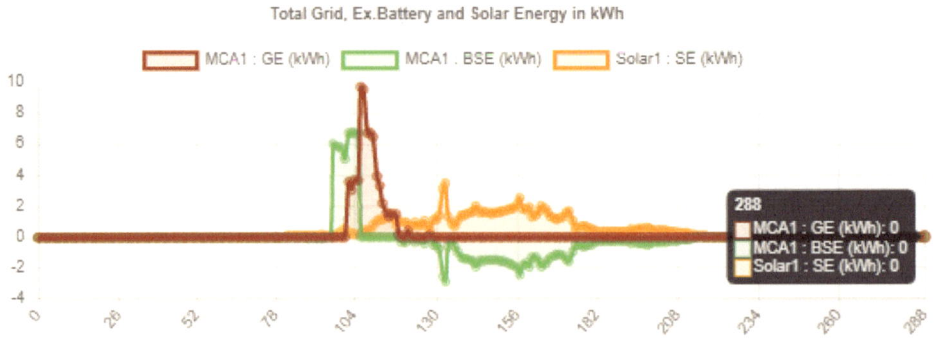

Figure 6.
Grid, station battery, and solar energy distribution for uncontrolled charging.

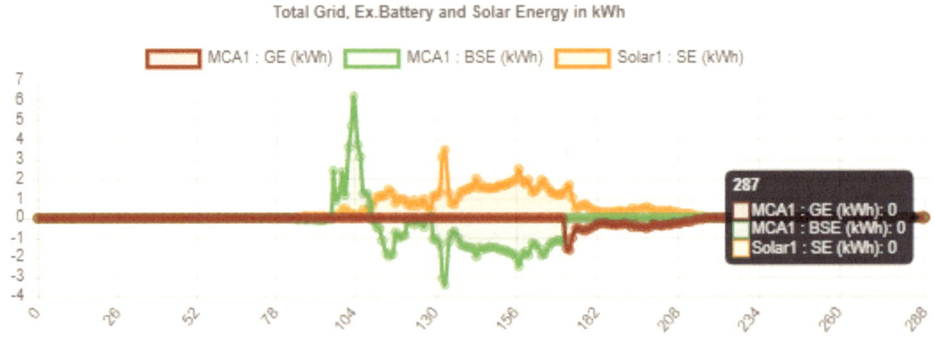

Figure 7.
Grid, station battery, and solar energy distribution for V2G charging.

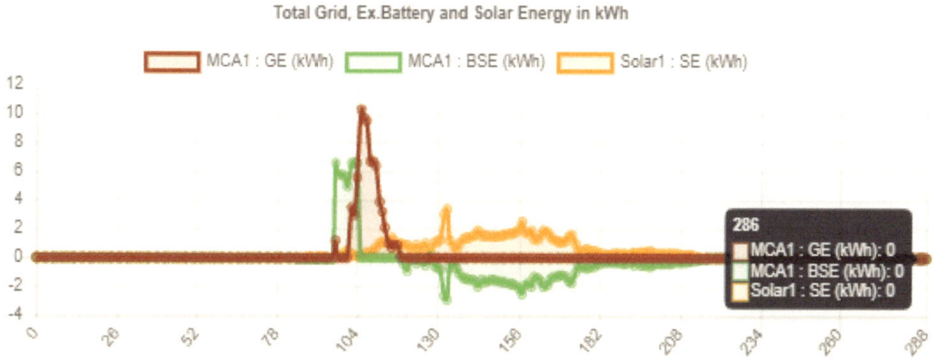

Figure 8.
Grid, station battery, and solar energy distribution for G2V charging.

Flat cost charging is used to simulate SOC-based charging; the three charging modes have the same unit cost. **Figures 6–8** show how uncontrolled charging and G2V result in the highest demand peak from the grid. The V2G has a lower peak because it allows only a few EVs to charge. According to our database, the majority of EVs are not far from our faculty. We also include simulation assumptions, such as when the EVs begin their journey with 100% SOC each day. When EVs arrive at the faculty, they have an average of more than 80% since they are not far away from the faculty. It causes them to inject excess energy into the grid as an energy source until it reaches 80% of SOC. It helps in reducing the grid's need for additional energy generation and the demand for power supply resources. **Figure 9** shows the simulation observation of G2V slow charging (**Figure 9**).

However, V2G and G2V permit obtaining energy consumption from energy resources and charging the vehicle when implementing controlled charging. When the SOC exceeds 50%, fast charging begins to preserve the lifetime of the battery from the full depth of discharge.

G2V controlled charging is combined with slow charging to improve DSB. It reduces grid dependability while improving DSB, as illustrated in **Figure 9**.

Figure 10 shows that DSB is calculated for all charging strategies based on SOC charging. EVs charge in average charging (6.6 kW) mode when uncontrolled charging and G2V charging mode are enabled. Both have almost the same DSB. In V2G, few

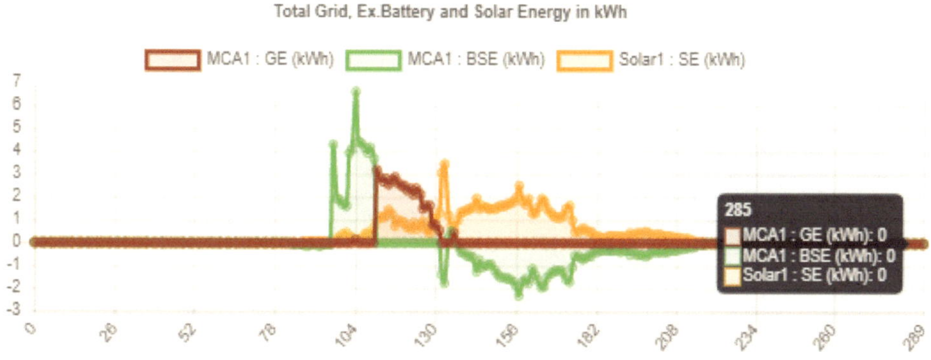

Figure 9.
Grid, station battery, and solar energy distribution for G2V combined with slow charging.

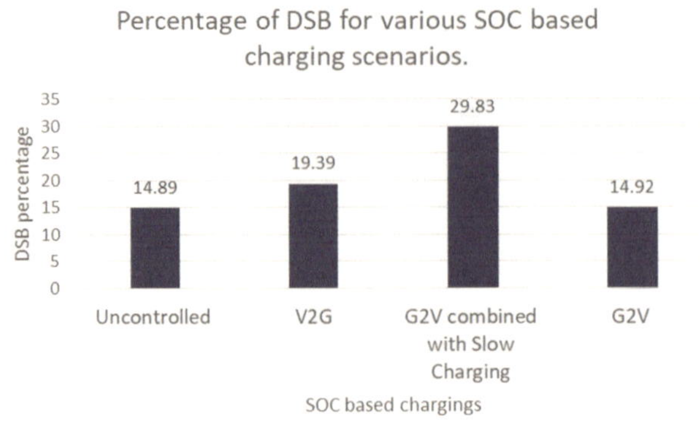

Figure 10.
DSB percentage for SOC-based charging scenarios.

cars need to be charged by the charging station, resulting in a moderate DSB value. When G2V is paired with slow charging, the DSB rises.

2.3.2 TOU-tariff-based charging

The TOU-based charging scenario simulates three different charging options. It includes slow, average, and fast charging. The charging powers are 2.3 kW, 6.6 kW, and 30 kW, respectively. In TOU-tariff-based charging, two major streams are considered: cost and direct solar benefit. The charging station's various start times are simulated to determine the most advantageous point of solar PV-based charging. The simulation observations are depicted in **Figures 11–13**.

This system is simulated with different station charging start times for the EV to see how PV energy affects EV charging. **Figure 14** summarizes the simulation results.

According to the findings, charging that begins at 11 a.m. has a higher DSB. However, our database shows that the few cars left the faculty before 11 a.m. Fast charging is assumed to have a 30-kW unit charge power excess for a few EVs in this simulation. As a result, the total demand to be achieved is reduced. Aside from that, the charging start time of 10 a.m. has a higher DSB. The slow and average charging have better DSB than the fast charging. The proportion of PV charging has increased, while the reliance on the public grid has decreased.

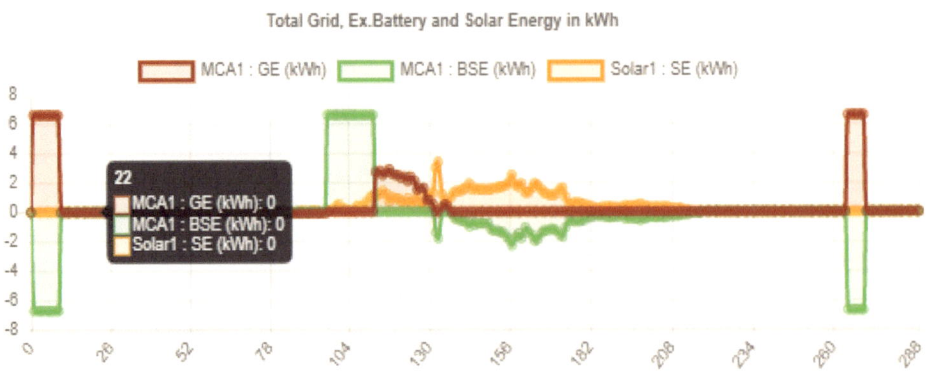

Figure 11.
Grid, station battery, and solar energy distribution for slow charging.

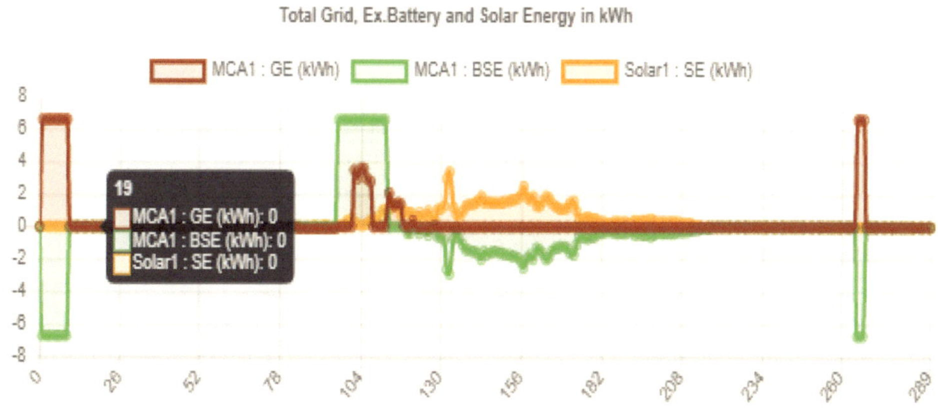

Figure 12.
Grid, station battery, and solar energy distribution for average charging.

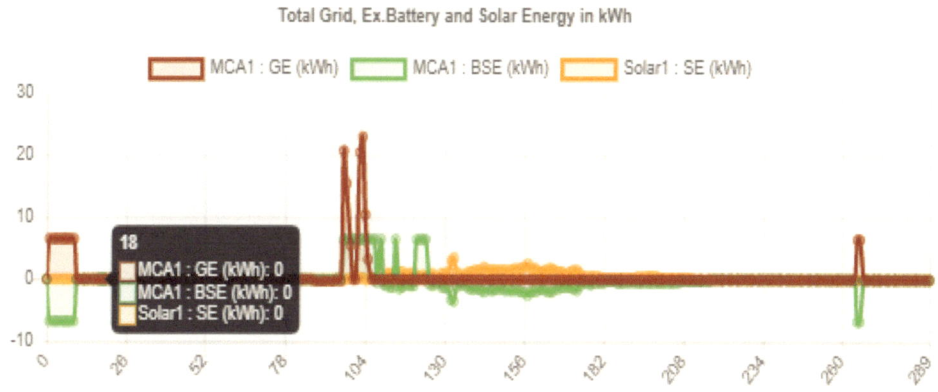

Figure 13.
Grid, station battery, and solar energy distribution for fast charging.

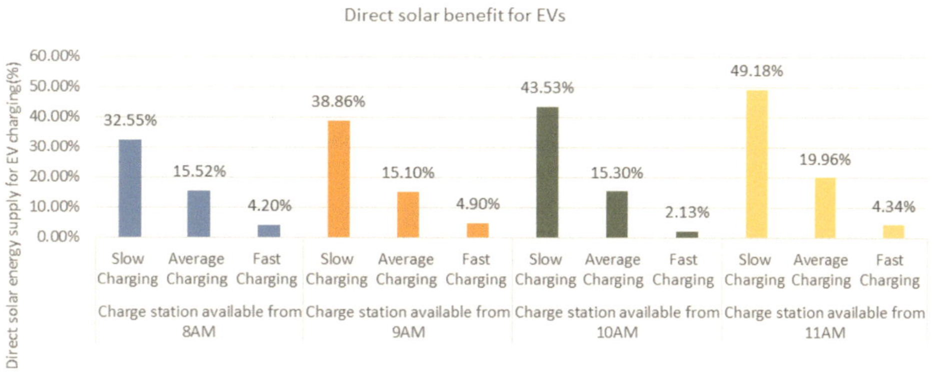

Figure 14.
The DSB results for EVs under TOU tariff.

Besides that, the stationary storage lasts longer, preventing rapid discharge. It does not exceed its minimum SOC (20%). The stationary storage energy decreases as it approaches its capacity limit with minimum SOC, and it is then supplied by the public

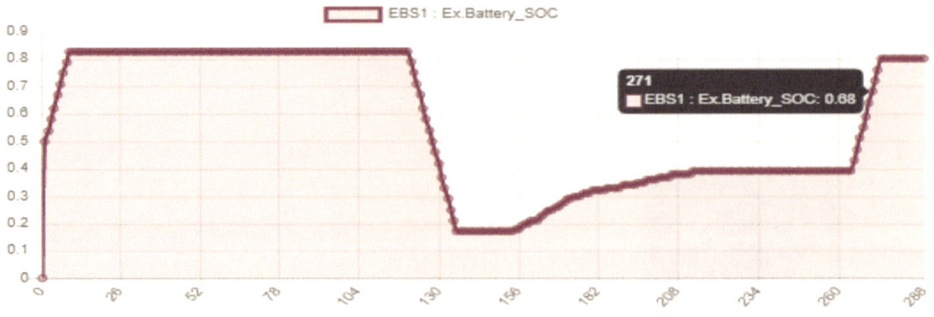

Figure 15.
Station battery SOC pattern for slow charging at 10 a.m.

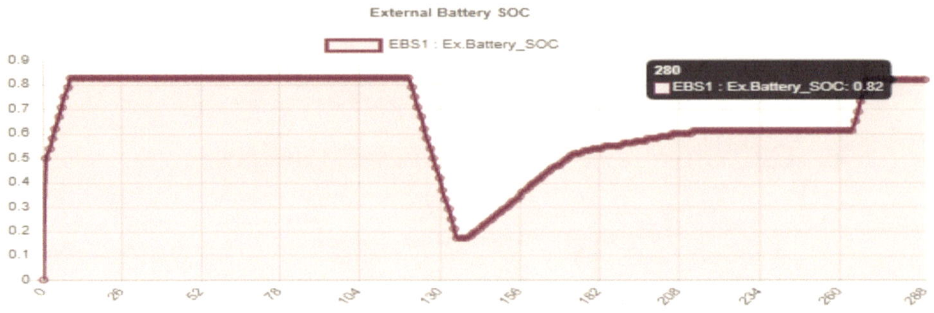

Figure 16.
Station battery SOC pattern for average charging at 10 a.m.

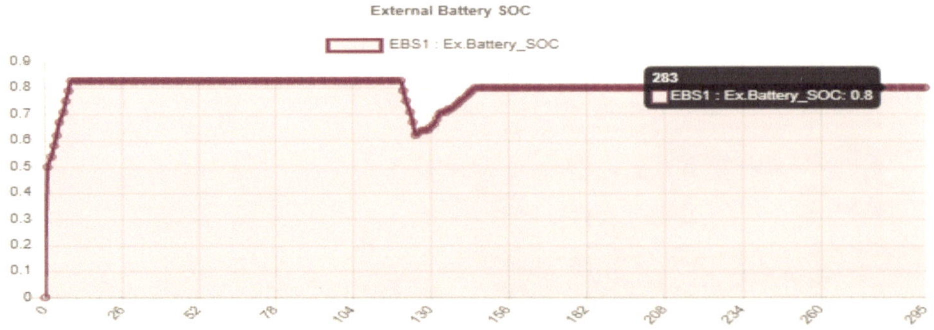

Figure 17.
Station battery SOC pattern for fast charging at 10 a.m.

grid during off-peak hours. The charge station battery SOC pattern is shown in **Figures 15–17** for slow, average, and fast charging, respectively.

As per the results, in slow and average charging modes, PV and stationary storage share more power. DSB is also increased when the charge station charging time is set to 10 a.m. in the university charging unit.

TOU tariff charging is expected to be cost-effective in PV-powered EV charging stations when compared with SOC-based charging. This creates a win-win situation for both charging station operators and EV owners. **Figures 18–20** show the graphical representation of the total and profit costs for both EV users and station owners.

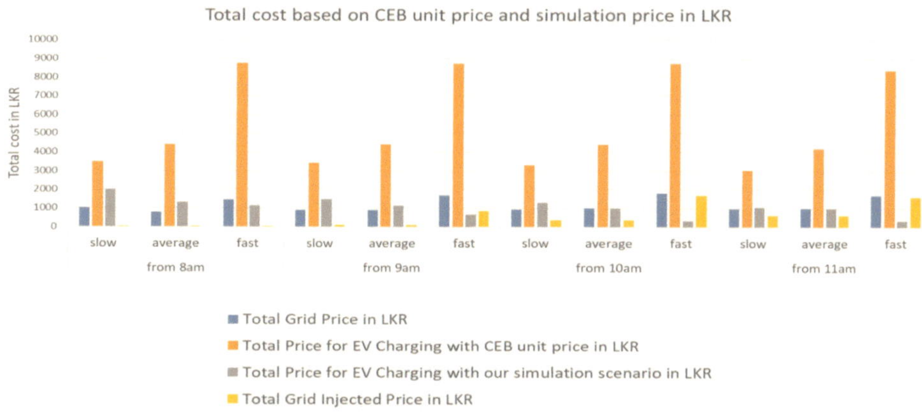

Figure 18.
Total cost based on CEB unit cost and simulation unit cost.

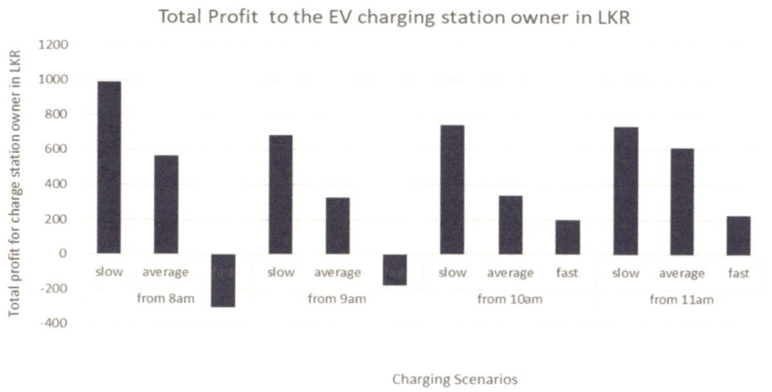

Figure 19.
Total profit for the EV owner in LKR (Sri Lankan Rupees).

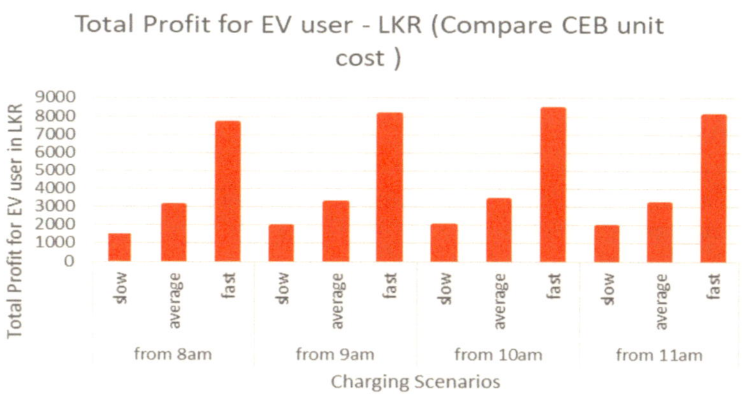

Figure 20.
Total profit for the charge station owner in LKR (Sri Lankan Rupees).

The cost benefit was calculated compared to the Ceylon Electricity Board's (CEB's) current EV charging price. As previously discussed, fast charging did not meet the total EV demand. Aside from that, slow and average charging has lower costs for both charge station operators and EV owners, which is more beneficial at 10 a.m. charging time. Fast charging also provides a higher grid-injected price benefit at the 10 a.m. start time to charge station operators than the other two start times.

3. Conclusion

There have been many exciting developments in ABM recently, and the use of truly adaptive agents within ABM is one promising guarantee that is still being explored. ABM is more than a simulation tool; it also aids in the reduction of operational risk and the development of new strategies for the organization. The incorporation of ML techniques into ABM should enable the creation of new and unique models.

Simulation model discussed focuses on the preliminary requirements and cost-effective model for charging in a university car park. Two main strategies were presented, SOC-based and TOU-tariff-based, which demonstrated improvements in terms of DSB and cost benefits. When compared to an uncontrolled charging strategy, SOC-based charging is safe and has many benefits. Uncontrolled EV charging causes a significant increase in power demand, which may cause power congestion or voltage issues in the power system. On the contrary, G2V charging with a slow charging mode has the advantage of distributing the charging load over time by limiting the peak power demand.

It is shown that the proposed system can effectively improve the DSB as well as cost benefits by implementing TOU-tariff-based charging. It is cost-effective for both charge station operators and EV owners. Two charging modes are advantageous for the requirements and feasibility conditions in the university car park: slow and average charging.

Our ABM charging station can communicate and collaborate with each agent to achieve the required system behavior. The MAS must be coordinated with its characteristics in order to attain the purpose [2, 3]. Scalability is an essential aspect to consider when creating practical MASs. The simulation model will expand the interaction between the agents without hesitation or delay as the number of EVs in the model expands.

Author details

Jaslin Shaleem Khan*, Malligama Arachchige Uditha Sudheera Navaratne and Janaka Bandara Ekanayake
Faculty of Engineering, Department of Electrical and Electronic Engineering, University of Peradeniya, Sri Lanka

*Address all correspondence to: jaslinshaleem93@gmail.com

References

[1] Julian V, Botti V. Multi-agent systems. Applied Science. 2019;**9**:1402. DOI: 10.3390/app9071402

[2] Oprea M. In: Reis R, editor. Applications of Multi-Agent Systems. Boston: Kluwer Academic Publishers; 2004. pp. 239-270

[3] Rocha J, Boavida-Portugal I, Gomes E. Introductory Chapter. London, UK: IntechOpen; 2017. DOI: 10.5772/intechopen.70241

[4] Hemapala KT, Jayasighe SL, Kulasekera AL. Multi-agent based control of smart grid. Communication Control Security Smart Grid. 2017;**1**: 321-345. DOI: 10.1049/PBPO095E_ch12

[5] Bodhinayake G, Gunawardena LHPN, Hemapala KTMU. Development of a multi agent system for voltage and outage monitoring. In: 2017 Third International Conference on Advances in Electrical, Electronics, Information, Communication and Bio-Informatics (AEEICB). Chennai, India: IEEE; 2017. pp. 156-162

[6] Kulasekera AL, Gopura RC, Hemapala KT, Perera N. A Review on Multi-agent Systems in Microgrid Applications. Kollam, India: IEEE; 2011. pp. 173-177. DOI: 10.1109/ISET-India.2011.6145377

[7] Logenthiran T, Srinivasan D, Shun TZ. Multi-Agent System for Demand Side Management in Smart Grid. Singapore: IEEE; 2011

[8] Et-Tolba EH, Maaroufi M, Ouassaid M. Demand side management in smart grid by multiagent systems technology. IEEE Access. 2014. DOI: 10.1109/ICMCS.2014.6911211

[9] Chuanchuan C, Jun Z, Panpan J. Multi-agent system applied to energy management system for renewable energy micro-grid. In: 2013 5th International Conference on Power Electronics Systems and Applications (PESA). Hong Kong, China: IEEE. Epub ahead of print December 2013. DOI: 10.1109/PESA.2013.6828222

[10] Aleksei Y, Kuzin GL, Demidova DV, Lukichev NA. Multi-agent System for Distributed Energy Microgrid: Simulation and Hardware-in-the-loop Physical Model. IEEE; 2020

[11] Li W, Logenthiran T, Woo WL, Phan V-T, Srinivasan D. Implementation of Demand Side Management of a Smart Home using Multi-agent System. IEEE; 2016

[12] Jiang W, Yang K, Mao R, Xue N, Zhuo Z. A multiagent-based hierarchical energy management strategy for maximization of renewable energy consumption in interconnected multi-microgrids. IEEE Access. 2019;**7**: 169931-169945. DOI: 10.1109/ACCESS.2019.2955552

[13] Ho TCT, Yu R, Lim JR, Franti P. Modelling implications & impacts of going green with EV in Singapore with multi-agent systems. In: Signal and Information Processing Association Annual Summit and Conference (APSIPA), 2014 Asia-Pacific. Chiang Mai, Thailand: IEEE; 2014. pp. 1–6

[14] Mocci S, Natale N, Ruggeri S, Pilo F. Multi-agent control system for increasing hosting capacity in active distribution networks with EV. In: 2014 IEEE Int. Energy Conf. ENERGYCON; Cavtat, Croatia. 2014. pp. 1409-1416

[15] Stifter M, Bermasser S. A multi-agent based approach for simulating G2V and V2G charging strategies for large electric vehicle fleets. In: 22nd Int. Conf. Exhib. Electr. Distrib. CIRED 2013; Stockholm, Sweden. 2013

[16] Ruggeri S, Pilo F, Natale N, Mocci S. Multi-agent control system to coordinate optimal electric vehicles charging and demand response actions in active distribution networks. In: Power Gener. Conf. RPG; Naples, Italy; 2014

[17] Lee R, Yazbeck S, Brown S. Validation and Application of Agent-based Electric Vehicle Charging Model. Elsevier; 2020. pp. 53-62

[18] Torres S, Barambones O, Gonzalez de Durana JM, Marzabal F, Kremers E, Wirges J. Agent-based modelling of electric vehicle driving and charging behavior. In: 2015 23rd Mediterr. Conf. Control Autom. MED, Torremolinos. 2015. pp. 459-464. DOI: 10.1109/MED.2015.7158791

[19] Chaudhari K, Kandasamy NK, Krishnan A, Ukil A, Gooi HB. Agent-based aggregated behavior modeling for electric vehicle charging load. IEEE Transactions on Industrial Informatics. 2019;**15**:856-868. DOI: 10.1109/TII.2018.2823321

[20] Wolbertus R, van den Hoed R. Expanding charging infrastructure for large scale introduction of electric vehicles. In: Electr. Veh. Symp, France. 2019

[21] Chaudhari KS. Agent-based Modelling of Electric Vehicle Charging for Optimized Charging Station Operation. Singapore: DR-NTU (Digital Repository of NTU). Nanyang Technological University; 2019

[22] van der Kam M, Peters A, van Sark W, Alkemade F. Agent-based modelling of charging behaviour of electric vehicle drivers. Journal of Artificial Societies and Social Simulations. 2019;**22**:7

[23] Lee S, Cherry J, Lee B, McDonald J, Safoutin M. HIL Development and Validation of Lithium-Ion Battery Packs. 2014. pp. 2014-01–1863. DOI: 10.4271/2014-01-1863

[24] Kazil J, Masad D, Crooks A. Utilizing python for agent-based modeling: The Mesa framework. In: Thomson R, Bisgin H, Dancy C, Hyder A, Hussain M, editors. Soc. Cult. Behav. Model. Cham: Springer International Publishing; 2020. pp. 308-317

[25] TOP 5 Significant Use Cases of Agent Based Modeling Simulation—Synergylabs. n.d. Available from: https://synlabs.io/top-5-significant-use-cases-of-agent-based-modeling-simulation/ [Accessed: May 25, 2022]

[26] Rand W. MACHINE LEARNING MEETS AGENT-BASED MODELING: WHEN NOT TO GO TO. 9. [Online]. Available from: https://www.semanticscholar.org/paper/MACHINE-LEARNING-MEETS-AGENT-BASED-MODELING-%3A-WHEN-Rand/65d565d5186345efc73457cf71cea61f6cfc8645

Chapter 6

Approximate Dynamic Programming: An Efficient Machine Learning Algorithm

Zhou Shaorui, Cai Ming and Zhuo Xiaopo

Abstract

We propose an efficient machine learning algorithm for two-stage stochastic programs. This machine learning algorithm is termed as projected stochastic hybrid learning algorithm, and consists of stochastic sub-gradient and piecewise linear approximation methods. We use the stochastic sub-gradient and sample information to update the piecewise linear approximation on the objective function. Then we introduce a projection step, which implemented the sub-gradient methods, to jump out from a local optimum, so that we can achieve a global optimum. By the innovative projection step, we show the convergent property of the algorithm for general two-stage stochastic programs. Furthermore, for the network recourse problem, our algorithm can drop the projection steps, but still maintains the convergence property. Thus, if we properly construct the initial piecewise linear functions, the pure piecewise linear approximation method is convergent for general two-stage stochastic programs. The proposed approximate dynamic programming algorithm overcomes the high dimensional state variables using methods from machine learning, and its logic capture the critical ability of the network structure to anticipate the impact of decisions now on the future. The optimization framework, which is carefully calibrated against historical performance, make it possible to introduce changes in the decisions and capture the collective intelligence of the experienced decisions. Computational results indicate that the algorithm exhibits rapid convergence.

Keywords: stochastic programming, piecewise linear approximation, machine learning, network, approximate dynamic programming

1. Introduction

Optimal learning addresses the challenge of how to collect information, as efficiently as possible, to make a decision in the present such that it minimizes the expectation of costs in the future with uncertainty. Collecting information is usually time consuming and expensive. For example, several large shippers, such as Amazon, Walmart, and IKEA, need to decide the quantity of products to ship from plants to warehouses to satisfy the retailers' demand. The retailer usually makes their decisions before knowing the real demand. Then, after they know the retail demand, they optimize the shipping plans between retailers and warehouses. The aforementioned problems generally can be treated as the two-stage stochastic programs. The decisions that the retailer made now (Stage 1) will determines the state while solving the problem in future (Stage 2). Therefore, an optimal decision can be made now if we can compute the expected cost function (the recourse function) of Stage 2. In this chapter, we propose an efficient machine learning algorithm that can collect information very efficiently based on the knowledge gradient and solve the problem optimally. The main gap between MAT and proposed algorithm is that our proposed can collect the information based on knowledge gradient and overcomes the "curse of dimensionality". Besides, it can transfer the problem into a polynomial solvable problem and has been proven convergent theoretically.

1.1 Motivation

Optimal learning is a rich filed that includes contributions from different communities. At the moment, this chapter focus on optimal learning in two-stage stochastic program, which is a practically important problem. Problems of this type arise in several areas in dynamic programming, in which the decision maker need to make temporal and spatial decisions before realizing events that will influence the decisions. For example, in empty container repositioning problems [1], shipping companies need to reposition empty containers before realizing the demand. In locomotive planning problems [2], railroads have to decide the schedule of trains in which locomotives are assigned before disruptions occur across the railway network. For relief distribution problems [3], humanitarian decision makers need to distribute emergency aid to disaster locations when the emergency aid materials are very scarce amidst great uncertainties. For job scheduling problems [4], the managers need to decide initial staffing levels and their working plans before the demand are realized. Most of the aforementioned applications are fully sequential problems, and they can be modeled as two-stage stochastic programming problems. Hence, the research of two-stage stochastic optimization in this chapter is very important. However, the main obstacle in most practical problem is that the expected cost function in Stage 2 is quite complex due to uncertainty. In this chapter, we propose a hybrid learning algorithm called *projected stochastic hybrid learning algorithm* (ProSHLA) to approximate the expected recourse function for two-stage stochastic programs. In order to demonstrate the efficiency of the algorithm, we also theoretically prove the convergence of the proposed algorithm mathematically.

In essence, ProSHLA is a hybrid of stochastic sub-gradient and piecewise linear approximation methods. The core of ProSHLA consists of a series of learning steps those provide information for updating the recourse function through a sequence of piecewise linear separable approximations, and a series of the projection steps those can guarantee convergence by implementing the stochastic sub-gradient method. The mathematical analysis and the computational results all demonstrates that when the initial piecewise linear approximation function is properly constructed for two-stage stochastic programs with network recourse, the learning algorithm can drop the projection steps without sacrificing convergence. Moreover, without the projection step, the learning algorithm only consists of a series of learning steps through a sequence of piecewise linear separable approximations, and can solve the practical complex problems very efficiently. Our innovative finding can help the practitioner and the scholar to understand the open problem that has puzzled them for decades: why does the piecewise linear approximation method can be efficient and convergent for stochastic programs with network recourse in practice. In this chapter, we provide the first theoretical support by our analytical results for the use of the piecewise linear approximation method in solving practical problems.

1.2 Literature review

In this chapter, we consider the two-stage stochastic programming problem as follows:

$$min\, c_0^T x + E_\omega[Q(x,\omega)] \tag{1}$$

s.t.,

$$Ax = b,$$
$$x \geq 0,$$

where $X \subset \mathfrak{R}^n$ denotes a convex compact set and the recourse function $Q(x,\omega)$ denotes the optimal value of the second stage problem:

$$Q(x,\omega) = \min c_1^T y(\omega) \tag{2}$$

s.t.,

$$W(\omega)y = h(\omega) - T(\omega)x,$$
$$y(\omega) \geq 0(\omega).$$

In the above model, variables x and y denote the decision variables of stage 1 and 2 problems, respectively. A, $W(\omega)$ are constraint matrices, and parameters c_0 and c_1 denote the first and second stage vectors of cost coefficients, respectively.

Stochastic programming models and solution methods has been examined by many researchers. Comprehensive reviews and discussions were performed by Wallace and Ziemba [5]. The expected recourse function is extremely complex to evaluate except for a few special cases. There are various approximation methods those can be categorized into four groups. Let $\widehat{Q}(x)$ denote the approximate function. The first group includes scenario methods which use the sample average of $Q(x,\omega_i)$ for several samples, $\omega_1, \omega_2, \ldots \omega_N$, to approximate the expected recourse function [6]. The approximation function is usually successively updated by the following function:

$$\widehat{Q}(x) = \frac{\sum_{i=1}^{N} Q(x,\omega_i)}{N}$$

Generally, the scenario method is very efficient, but it cannot guarantee to obtain the convergent solution.

The second group consists stochastic gradient techniques [7, 8], which updates solutions by using stochastic sub-gradients as directions. Usually, the approximate function can be successively updated by the following function:

$$\widehat{Q}(x) = \left(\bar{g}^k\right)^T x \tag{3}$$

where $\bar{g}^k$ denotes a smoothed estimate of the gradient of the expected recourse function at x for iteration k. This method can be proven convergent by projection [9] or recursive linearization [10], although the drawback of this method is that it is time-consuming.

The third group mainly consists of primal and dual decomposition methods. The use of primal and dual decomposition methods dates back to Benders decomposition [11]. Van Slyke and Wets [12] first adopted the L-shaped algorithm into the application of Benders decomposition to two-stage stochastic programs. Pereira and Pinto [13] proposed the stochastic dual dynamic programming (SDDP) method, which has been widely applied in many areas. SDDP uses Benders cuts to compute an outer approximation of a (convex) recourse function, and constructs feasible dynamic programming policies. SDDP has led to numerous related approximation methods those are based on the same logic but seek to improve the approximation procedures by exploiting the underlying structure of the particular applications. These methods consist of use of inexact cuts [14], risk-averse variants [15], embedding SDDP in the scenario tree framework [16]. The convergence of SDDP and related methods has been proven by [17], for linear programs by Girardeau et al. [18].

The fourth group includes separable approximation methods [19, 20]. This type of methods usually replaces the expected recourse function in Eq. (1) with separable approximation functions as follows:

$$\widehat{Q}(x) = \sum_{i=1}^{I} \widehat{Q}_i(x) \tag{4}$$

If the separable functions $\widehat{Q}_i(x)$ are piecewise linear or linear, we can replace the expected recourse function in Eq. (1) with $\widehat{Q}(x)$. Then we can solve the problem as a pure network flow problem for network recourse problems, which is polynomial solvable. Thus, it is very efficient. For example, Godfrey and Powell [21] proposed an adaptive piecewise concave approximation (CAVE) algorithm, and the experimental performance of the algorithm shows exceptionally good. However, there was none provable convergent results in their study. In order to provide convergent solutions, Cheung and Powell [19] proposed an approximation algorithm (SHAPE), which uses sequences of strongly convex approximation functions. However, the strongly convex functions require to construct a nonlinear term, and the strongly convex term might damage the pure network structure and need additional computational effort. This chapter intends to introduce an accurate and efficient approximations with the convergence property.

1.3 Contributions of the algorithms

In this chapter, we aim to develop a convergent method that can efficiently approximate the expected recourse function for two-stage stochastic programs. The main contributions are listed as following:

1. We propose a new convergent hybrid learning algorithm to approximate the expected recourse function for two-stage stochastic programs.

2. Through rigorous mathematical analysis, we prove the convergence of the proposed algorithm for general two-stage stochastic programs. The

computational results and mathematical analysis both reveals that the algorithm can drop the projection step without sacrificing the convergence for two-stage stochastic programs with network recourse if we can properly construct the initial piecewise linear approximation functions. That means that a pure piecewise linear approximation can be indeed convergent, which is highly consistent with industry practices. This interesting finding answers the open question which has puzzled scholars for more than a decade: why does the piecewise linear approximation work well for two-stage stochastic programs in industry? Our mathematical analysis can provide the first theoretical support.

3. A series of performance analysis has been conducted. The computational results reveal the efficiency of the proposed algorithms and the proposed algorithms are distribution-free. Furthermore, the convergence rate can be affected by the granularity of the initial function (δ). Small granularity usually leads to a high convergence rate. Finally, the computational results also show that the proposed algorithm is very competitive for high dimensional problems.

4. Compared with MAT, the proposed algorithm can collect the information based on knowledge gradients and use it to update the recourse function by learning steps. It can overcome the "curse of dimensionality". Moreover, it can transfer the problem into a polynomial solvable problem.

The remainder of this chapter is organized as follows. Section 2 presents the description and convergence analysis of the algorithm for general two-stage stochastic programs. The algorithm (without projection steps) for two-stage stochastic programs with network recourse are shown in Section 3. Section 4 demonstrates computational experiments based on an application of the empty container repositioning problem. Section 5 presents the conclusions and outline directions for future research.

2. Description and convergence analysis of ProSHLA for general two-stage stochastic programs

In this section, ProSHLA is first introduced. Subsequently, we analyze the convergence of ProSHLA for general two-stage stochastic programs.

2.1 Description of ProSHLA

To present ProSHLA mathematically, we let, at each iteration k,

$\alpha_k =$ (possibly random) positive step size;

$\overline{Q}(x) =$ expected recourse function, that is, $E_\omega[Q(x, \omega)]$;

$\widehat{Q}^k(x) =$ a convex differentiable approximation of $\overline{Q}(x)$;

$\widehat{q}^k(x) =$ a subgradient of $\widehat{Q}^k(x)$ at x, that is $\widehat{q}^k(x) \in \partial\widehat{Q}^k(x)$;

$\overline{g}^k =$ a smoothed estimate of the gradient of $\overline{Q}(x)$ at iteration k;

$g^k =$ a stochastic subgradient of $\overline{Q}(x)$ at x^k, that is, $g^k \in \partial Q(x^k, \omega^{k+1})$;

$H_k = \{\omega_1, \omega_2, \dots \omega_N\} =$ the history up to (and including) iteration k.

For a general non-smooth convex function $\widehat{Q}(x)$, its sub-differential can be defined as follows:

$$\partial\widehat{Q}(x) = \left\{\widehat{q}(x) \in \Re^n : \widehat{Q}(y) - \widehat{Q}(x) \geq \widehat{q}(x)^{\mathrm{T}} (y - x)\right\}.$$

We combine Eqs. (1), (3), and (4) to form an approximation at iteration k as follows:

$$min\, c_0^T x + \widehat{Q}^0(x) + \left(\overline{g}^k\right)^T x \tag{5}$$

In this study, we approximate the expected recourse function at iteration k via a convex, differentiable approximation $\widehat{Q}^0(x)$ with a linear correction term $\left(\overline{g}^k\right)^T x$. At each iteration, the linear correction term $\left(\overline{g}^k\right)^T x$ are introduced to improve the initial approximation $\widehat{Q}^0(x)$. Note that here we use a convex initial approximation function $\widehat{Q}^0(x)$, whereas SHAPE uses strongly convex approximation functions. SHAPE will introduce a nonlinear term in the approximation function to maintain the strong convexity property, and it might destroy the pure network flow problem structure and demands additional computational effort. Moreover, we do not calculate $\overline{g}^k$ in the usual manner to obtain stochastic sub-gradients in this study. We use the following form in our model instead:

$$min\, c_0^T x + \widehat{Q}^k(x) + \alpha_k \left(g^k - \widehat{q}^k(x)\right)^T x, \tag{6}$$

where $\widehat{Q}^k(x)$ is updated as follows:

$$\widehat{Q}^{k+1}(x) = \widehat{Q}^k(x) + \alpha_k \left(g^k - \widehat{q}^k(x)\right)^T x \tag{7}$$

The greatest merit of updating $\widehat{Q}^{k+1}(x)$ in the above way is that it can retain the stochastic sub-gradients $\left(\widehat{q}^k(x^k), \widehat{q}^{k-1}(x^{k-1}), \ldots, \widehat{q}^0(x^0)\right)$ used in the previous iterations. Thus, in iteration k, the objective function involves a weighted average of stochastic sub-gradients in the past $(k-1)$ iterations. As shown later in Lemma 2, $\overline{g}^k$ in Eq. (5) is a linear combination of $g^1, g^2, \ldots g^{k-1}$.

Let $P_X : R^n \to X$ be the orthogonal projection onto X [9]. Then, we can obtain a sequence of solutions $\{x^k\}$ using the following procedure (**Figure 1**).

Generally, ProSHLA consists of two-level loops. In the first-level loop, there exists a series of *passes*, and in the second-level loop, the exists a series of projection steps, which include the step 5 and 6. We first construct an initial bounded and piecewise linear convex approximation function $\widehat{Q}^0(x)$ at the beginning of the first pass, then the initial solution x^0 can be obtained by solving problem (1). A realization of the random quantity $\omega \in \Omega$ can be drawn, and then we can obtain a stochastic sub-gradient of $\overline{Q}^0(x)$ by solving the resulting deterministic problem. Compared with the slope of $\widehat{Q}^0(x)$ and the stochastic sub-gradient at $x = x^0$, the difference of these two slopes can be used as a linear term to update $\widehat{Q}^0(x)$. Subsequently, we can obtain a

ProSHLA

Step 1. Let the pass counter $m = 0$ and the iteration counter $k = 0$. Construct an initial *convex* function $\hat{Q}^0\ (x)$. Let $\bar{m}$ be the number of the first iteration of the m pass. Maintain a sequence of points: $\{x^k\}$.

Step 2. Obtain $x^0 = \arg\min\ \hat{Q}^0\ (x)$.

Step 3. Obtain $\hat{q}^k(x^k)$ and g^k. Let $\bar{m} = k$, $x^{\bar{m}} = x^k$, and $\hat{q}^{\bar{m}}(x^{\bar{m}}) = \hat{q}^k(x^k)$. $\hat{Q}^0\ (x)$ can be updated by

$$\hat{Q}^{k+1}(x) = \hat{Q}^k(x) + \alpha_k(g^k - \hat{q}^k(x))^T x.$$

Step 4. Obtain x^{k+1} by

$$x^{k+1} = \arg\min\ \hat{Q}^{k+1}(x). \tag{8}$$

Step 5. If $|\hat{q}^{\bar{m}}(x^{k+1}) - \hat{q}^{\bar{m}}(x^{\bar{m}})| > 0$ or $x^{k+1} = x^{\bar{m}}$, then update the pass counter $m = m + 1$ and go to step 7; otherwise, go to Step 6.

Step 6. Obtain x^{k+1} by

$$x^{k+1} = P_X(x^k - \alpha_k g^k). \tag{9}$$

Thereafter, go to Step 5.

Step 7. Check for convergence (e.g. an improvement in $\hat{Q}^k$ in the last K iterations). If the check fails, set $k = k + 1$ and go to Step 3; otherwise, terminate.

Figure 1.
The procedure of ProSHLA.

new solution x^{k+1} using the updated approximation function. If the sub-gradient vectors $\hat{q}^{\bar{m}}(x^{k+1})$ of the newly obtained solution x^{k+1} are equal to sub-gradient of solution $x^{\bar{m}}$, that is $\hat{q}^{\bar{m}}(x^{\bar{m}})$, the piecewise linear approximation might have jumped into a local optimum. Subsequently, ProSHLA need to jump out from local optimum by implementing projection steps in the second-level loop. If we obtain a new solution x^{k+1} in the second-level loop and the sub-gradient $\hat{q}^{\bar{m}}(x^{k+1})$ is different from sub-gradient $\hat{q}^{\bar{m}}(x^{\bar{m}})$, then ProSHLA can jump out the second-level loop, and comes to the end of second pass. Thus then, we can repeat the entire process. Finally, ProSHLA will be terminated when the total absolute change in $\hat{Q}^l(x)$ over a certain number of iterations is low (e.g. $\sum_{l=k-M+1}^{k} \|\hat{Q}^l(x) - \hat{Q}^{l-1}(x)| < \delta$).

Here we point out the main difference between SHAPE and ProSHLA. The most remarkable difference is that ProSHLA uses convex approximation functions while SHAPE uses strongly convex approximation functions. The strongly convexity always maintains a nonlinear term in the approximation function. And this term might destroy the pure network flow structure and causes additional computational effort. To overcome the drawback of the SHAPE, we introduce the projection step in the second-level loop and construct approximation functions. Particularly, the approximation functions in the ProSHLA is NOT strictly convex, while it needs to be strictly convex in SHAPE. Without the projection step in the second-level loop, the ProSHLA might stuck in the corner local –optimum for stochastic linear programs. Thus, ProSHLA can work well for most practical stochastic linear programs, because most of practical stochastic programs are piecewise convex problems.

2.2 Convergence analysis of ProSHLA

Firstly, we demonstrate the convergence theorem of ProSHLA in this subsection. Then, several properties of approximation are presented. Finally, we use these properties to prove the convergence of ProSHLA.

Without loss of generality, the following assumptions are listed.

(A.1) $X \subset \mathfrak{R}^n$ is compact and convex.

(A.2) $E_\omega Q(x_1, \omega)$ is convex, finite and continuous on X.

(A.3) g^k is bounded such that $\|g^k\| \leq c_1$ for each $\omega \in \Omega; \widehat{q}^k$ is bounded such that $\|\widehat{q}^k\| \leq c_2$ for each $\omega \in \Omega$.

(A.4) Piecewise linear function $\widehat{Q}^k(x)$ are convex, implying that

$$\widehat{Q}^k(x_1) - \widehat{Q}^k(y_1) \leq \widehat{q}^k(x_1)^T(x_1 - y_1).$$

(A.5) The stepsize α_k are $\mathcal{H}_k$ measurable and satisfy

$$0 < \alpha_k < 1, \sum_{k=0}^{\infty} E\{\alpha_k^2\} \leq \infty$$

Except for the assumption from (A.1) to (A.5), we also introduce the following assumption to characterize the piecewise linear convex approximation functions.

(A.6) There exists a positive b and a constant δ, such that for any two points $x_1, y_1 \in X$, if$|x_1 - y_1| > \delta$, then $|\widehat{q}(x_1) - \widehat{q}(y_1)| \geq b|x_1 - y_1|$. If there exists $\widehat{q}(x_1)$ and $\widehat{q}(y_1)$ such that $\widehat{q}^k(x_1) - \widehat{q}^k(y_1) = 0$, then $|x_1 - y_1| \leq \delta$. If $\delta \to 0$, then the function corresponds to a strongly convex function, if $\delta \to \infty$, then the function becomes purely linear.

Given assumption (A.1)–(A.6), we obtain the following theorem of ProSHLA.

Theorem 1. If assumptions (A.1)–(A.6) are satisfied, then the sequence of $\{x_1^k\}$ generated by algorithm ProSHLA converges almost surely to the optimal solution $x_1^* \in X^*$ of problem (1).

In order to prove the Theorem 1, we need to use the following Martingale convergence theorem and three lemmas.

Martingale Convergence Theorem. A sequence of random variables $\{W^k\}$, which are $\mathcal{H}_k$ measurable, is said to be a super-martingale if there exists the sequence of conditional expectations $E\{\{W^{k+1}|\mathcal{H}_k\}$ and satisfies $E\{\{W^{k+1}|\mathcal{H}_k\} \leq W^k$.

Theorem 2. (From reference [22]) Let W^k be a positive super-martingale. Then, W^k converges to a finite random variables a.s.

From the above theorem, we can conclude that W^k is a stochastic decreasing analogue essentially.

Based on the convexity property, the optimal solution for problem (8) at iteration $\overline{m}$ can be characterized by the following inequality:

$$\left(\widehat{q}^{\overline{m}}(x_1^{\overline{m}})\right)^T(x_1 - x_1^{\overline{m}}) \geq 0, \quad \forall x_1 \in X \tag{8}$$

To obtain Theorem 1, the following three lemmas are required. The first lemma shows that the difference between the solutions of two consecutive update processes will be bounded by the step-size and the stochastic gradient. The second lemma indicates that the approximation $\widehat{Q}^k(x_1)$ is finite. The third lemma shows that T^k (which will be denoted in Lemma 3) is bounded.

Lemma 1. For any two iterations $j \in \left[\overline{m+1}, \overline{m+2}\right), i \in \left[\overline{m}, \overline{m+1}\right)$, solutions x_1^j and x_1^i obtained by ProSHLA can be characterized by the following inequality:

$$\alpha_{\overline{m}} g^{\overline{m}} \left(x_1^i - x_1^j\right) \leq (\alpha_{\overline{m}} c_1)^2 / b \tag{9}$$

Proof. Consider a special case, where i and j corresponds to two consecutive iterations. Let $i = \overline{m+1} - 1$ and $j = \overline{m+1}$. Based on (10), we can obtain that,

$$\left(\widehat{q}^{\overline{m+1}}\left(x_1^{\overline{m+1}}\right)\right)^T \left(x_1 - x_1^{\overline{m+1}}\right) \geq 0, \forall x_1 \in X \tag{10}$$

According to the approximation function's updating rule, we can conclude that,

$$\left(\widehat{q}^{\overline{m+1}-1}\left(x_1^{\overline{m+1}}\right) + \alpha_{\overline{m+1}-1}\left(g^{\overline{m+1}-1} - \widehat{q}^{\overline{m+1}-1}\left(x_1^{\overline{m+1}-1}\right)\right)\right)^T \left(x_1 - x_1^{\overline{m+1}}\right) \geq 0, \forall x_1 \in X \tag{11}$$

Substituting x_1 with $x_1^{\overline{m+1}-1}$ in Eq. (13), we can obtain

$$\begin{aligned} &\alpha_{\overline{m+1}-1}\left(g^{\overline{m+1}-1} - \widehat{q}^{\overline{m+1}-1}\left(x_1^{\overline{m+1}-1}\right)\right)^T \left(x_1^{\overline{m+1}-1} - x_1^{\overline{m+1}}\right) \\ &\geq \widehat{q}^{\overline{m+1}-1}\left(x_1^{\overline{m+1}}\right)^T \left(x_1^{\overline{m+1}} - x_1^{\overline{m+1}-1}\right) \end{aligned} \tag{12}$$

If we arrange the above terms, then we can obtain the inequality below:

$$\begin{aligned} &\alpha_{\overline{m+1}-1}\left(g^{\overline{m+1}-1}\right)^T \left(x_1^{\overline{m+1}-1} - x_1^{\overline{m+1}}\right) \\ &\geq \widehat{q}^{\overline{m+1}-1}\left(x_1^{\overline{m+1}}\right)^T \left(x_1^{\overline{m+1}} - x_1^{\overline{m+1}-1}\right) - \alpha_{\overline{m+1}-1}\left(\widehat{q}^{\overline{m+1}-1}\left(x_1^{\overline{m+1}-1}\right)\right)^T \left(x_1^{\overline{m+1}} - x_1^{\overline{m+1}-1}\right) \\ &= \left(\widehat{q}^{\overline{m+1}-1}\left(x_1^{\overline{m+1}}\right) - \widehat{q}^{\overline{m+1}-1}\left(x_1^{\overline{m+1}-1}\right)\right)^T \left(x_1^{\overline{m+1}} - x_1^{\overline{m+1}-1}\right) \\ &\quad + \left(1 - \alpha_{\overline{m+1}-1}\right)\left(\widehat{q}^{\overline{m+1}-1}\left(x_1^{\overline{m+1}-1}\right)\right)^T \left(x_1^{\overline{m+1}} - x_1^{\overline{m+1}-1}\right) \end{aligned} \tag{13}$$

When iteration $\overline{m+1} - 1$ and $\overline{m+1}$ are not in the same update process, then it means that $\widehat{q}^{\overline{m+1}-1}\left(x_1^{\overline{m+1}-1}\right) \neq \widehat{q}^{\overline{m+1}-1}\left(x_1^{\overline{m+1}}\right)$. According to assumption (A.6), we can conclude that $|\widehat{q}^{\overline{m+1}-1}\left(x_1^{\overline{m+1}-1}\right) - \widehat{q}^{\overline{m+1}-1}\left(x_1^{\overline{m+1}}\right)| \geq b|x_1^{\overline{m+1}-1} - x_1^{\overline{m+1}}|$.

According to Eqs. (10) and (13), and $0 < \alpha_{\overline{m+1}-1} < 1$, we can obtain

$$\begin{aligned} &\alpha_{\overline{m+1}-1}\left(g^{\overline{m+1}-1}\right)^T \left(x_1^{\overline{m+1}-1} - x_1^{\overline{m+1}}\right) \\ &\geq b\left\|x_1^{\overline{m+1}-1} - x_1^{\overline{m+1}}\right\|^2 + \left(1 - \alpha_{\overline{m+1}-1}\right)\left(\widehat{q}^{\overline{m+1}-1}\left(x_1^{\overline{m+1}-1}\right)\right)^T \left(x_1^{\overline{m+1}} - x_1^{\overline{m+1}-1}\right) \\ &\geq b\left\|x_1^{\overline{m+1}-1} - x_1^{\overline{m+1}}\right\|^2. \end{aligned}$$

Applying Schwartz' inequality, we can get the inequality below:

$$\alpha_{\overline{m+1}-1}\left\|g^{\overline{m+1}-1}\right\|\bullet\left\|x_1^{\overline{m+1}-1}-x_1^{\overline{m+1}}\right\|\geq\alpha_{\overline{m+1}-1}\left(g^{\overline{m+1}-1}\right)^T\left(x_1^{\overline{m+1}-1}-x_1^{\overline{m+1}}\right)$$
$$\geq b\left\|x_1^{\overline{m+1}-1}-x_1^{\overline{m+1}}\right\|^2$$

Therefore, $\left\|x_1^{\overline{m+1}-1}-x_1^{\overline{m+1}}\right\|\leq\alpha_{\overline{m+1}-1}\bullet c_1/b$. We can obtain the inequality below:

$$\alpha_{\overline{m+1}-1}\left(g^{\overline{m+1}-1}\right)^T\left(x_1^{\overline{m+1}-1}-x_1^{\overline{m+1}}\right)\leq(\alpha_{\overline{m+1}-1}c_1)^2/b \tag{14}$$

For any $i\in[\overline{m},\overline{m+1}-1]$, $g^i=g^{\overline{m}}=g^{\overline{m+1}-1}$ and $\widehat{q}^i(x_1)=\widehat{q}^{\overline{m}}(x_1)=\widehat{q}^{\overline{m+1}-1}(x_1)$. Thus,

$$\alpha_{\overline{m}}\left(g^{\overline{m}}\right)^T\left(x_1^i-x_1^{\overline{m+1}}\right)\leq(\alpha_{\overline{m}}c_1)^2/b \tag{15}$$

For any $i\in[\overline{m},\overline{m+1}-1]$ and $j\in[\overline{m+1},\overline{m+2}-1]$, $g^j=g^{\overline{m+1}}=g^{\overline{m+2}-1}$ for any $x_1\in X$. Thus,

$$\alpha_{\overline{m}}\left(g^{\overline{m}}\right)^T\left(x_1^i-x_1^j\right)\leq(\alpha_{\overline{m}}c_1)^2/b \tag{16}$$

□

Lemma 2. In iteration k, the approximation function $\widehat{Q}^k(x_1)$ can be written as $\widehat{Q}^k(x_1)=\widehat{Q}^0(x_1)+\left(\overline{g}^k\right)^Tx_1$, where $\overline{g}^k$ is a finite vector.

Proof. According to Eq. (5) in Proposition 1, we can conclude that $\overline{g}^{k+1}$ is a linear combination of $g^1,g^2,\ldots,g^k$. Since g^k and $\widehat{q}^0(x_1)$ are finite, there will exists a finite and positive vector $\widehat{d}$ such that

$$\widehat{d}\geq\max_k\left|g^k-\widehat{q}^0\left(x_1^k\right)\right| \tag{17}$$

According to Lemma 2 in [19], we can conclude that $\overline{g}^{k+1}\geq\widehat{d}$. □

Let $T^k=\widehat{Q}^k(x_1^*)-\widehat{Q}^k(x_1^k)$, where x_1^* represents the optimal solution. The following Lemma characterizes the difference between T^{k+1} and T^k.

Lemma 3. For any two iterations $i\in[\overline{m-1},\overline{m}-1]$ and $j\in[\overline{m},\overline{m+1}-1]$, T^i and T^j satisfy

$$T^j-T^i\leq\alpha_{\overline{m}}\left(g^{\overline{m}}\right)^T\left(x_1^i-x_1^j\right)+\alpha_{\overline{m}}\left(g^{\overline{m}}\right)^T\left(x_1^*-x_1^i\right). \tag{18}$$

Proof. The special case is first considered. Let $i=\overline{m}$ and $j=\overline{m+1}$. By re-writing $x_1^*-x^{\overline{m+1}}$ as $x_1^*-x^{\overline{m}}+x^{\overline{m}}-x^{\overline{m+1}}$, we can obtain the following equation:

$$\widehat{Q}^{\overline{m+1}}(x_1)=\widehat{Q}^{\overline{m+1}-1}(x_1)+\alpha_{\overline{m+1}-1}\left(g^{\overline{m+1}-1}-\widehat{q}^{\overline{m+1}-1}\left(x_1^{\overline{m+1}-1}\right)\right)^Tx_1$$
$$=\widehat{Q}^{\overline{m}}(x_1)+\alpha_{\overline{m}}\left(g^{\overline{m}}-\widehat{q}^{\overline{m}}\left(x_1^{\overline{m}}\right)\right)^Tx_1$$

Then,

$$\begin{aligned}
T^{\overline{m+1}} - T^{\overline{m}} &= \widehat{Q}^{\overline{m}}\left(x_1^*\right) + \alpha_{\overline{m}}\left(g^{\overline{m}} - \widehat{q}^{\overline{m}}\left(x_1^{\overline{m}}\right)\right)^T x_1^* - \left(\widehat{Q}^{\overline{m}}\left(x_1^{\overline{m+1}}\right)\right. \\
&\quad \left. + \alpha_{\overline{m}}\left(g^{\overline{m}} - \widehat{q}^{\overline{m}}\left(x_1^{\overline{m}}\right)\right)^T x_1^{\overline{m+1}}\right) - \left(\widehat{Q}^{\overline{m}}\left(x_1^*\right) - \widehat{Q}^{\overline{m}}\left(x_1^{\overline{m}}\right)\right) \\
&= \alpha_{\overline{m}}\left(g^{\overline{m}} - \widehat{q}^{\overline{m}}\right)^T\left(x_1^* - x_1^{\overline{m}} + x_1^{\overline{m}} - x_1^{\overline{m+1}}\right) + \widehat{Q}^{\overline{m}}\left(x_1^{\overline{m}}\right) - \widehat{Q}^{\overline{m}}\left(x_1^{\overline{m+1}}\right) \\
&= \underbrace{\left(\widehat{Q}^{\overline{m}}\left(x_1^{\overline{m}}\right) - \widehat{Q}^{\overline{m}}\left(x_1^{\overline{m+1}}\right) - \alpha_{\overline{m}}\left(\widehat{q}^{\overline{m}}\right)^T\left(x_1^{\overline{m}} - x_1^{\overline{m+1}}\right)\right)}_{I} \\
&\quad \underbrace{- \alpha_{\overline{m}}\left(\widehat{q}^{\overline{m}}\right)^T\left(x_1^* - x_1^{\overline{m}}\right)}_{II} + \underbrace{\alpha_{\overline{m}}\left(g^{\overline{m}}\right)^T\left(x_1^{\overline{m}} - x_1^{\overline{m+1}}\right)}_{III} + \underbrace{\alpha_{\overline{m}}\left(g^{\overline{m}}\right)^T\left(x_1^* - x_1^{\overline{m}}\right)}_{IV}
\end{aligned}$$

Considering each part individually, given that $\widehat{q}^{\overline{m}} \in \partial\widehat{Q}^{\overline{m}}\left(x_1^{\overline{m}}\right)$, by convexity of $\widehat{Q}^{\overline{m}}\left(x_1^{\overline{m}}\right)$, we can obtain

$$\widehat{Q}^{\overline{m}}\left(x_1^{\overline{m}}\right) - \widehat{Q}^{\overline{m}}\left(x_1^{\overline{m+1}}\right) \le \left(\widehat{q}^{\overline{m}}\right)^T\left(x_1^{\overline{m}} - x_1^{\overline{m+1}}\right) \tag{19}$$

Thus, the following expression is applicable.

$$\begin{aligned}
\widehat{Q}^{\overline{m}}\left(x_1^{\overline{m}}\right) - \widehat{Q}^{\overline{m}}\left(x_1^{\overline{m+1}}\right) &\le \left(\widehat{q}^{\overline{m}}\right)^T\left(x_1^{\overline{m}} - x_1^{\overline{m+1}}\right) \\
&= \left(1 - \alpha_{\overline{m}}\right)\left(\widehat{q}^{\overline{m}}\right)^T\left(x_1^{\overline{m}} - x_1^{\overline{m+1}}\right) + \alpha_{\overline{m}}\left(\widehat{q}^{\overline{m}}\right)^T\left(x_1^{\overline{m}} - x_1^{\overline{m+1}}\right)
\end{aligned} \tag{20}$$

Given Eq. (10) and $0 < \alpha_{\overline{m}} < 1$, we know that $(I) \le 0$. Additionally, from Eq. (10) and $0 < \alpha_{\overline{m}} < 1$, we know that $(II) \ge 0$.
Thus, $T^{\overline{m+1}} - T^{\overline{m}} \le \alpha_{\overline{m}}\left(g^{\overline{m}}\right)^T\left(x_1^{\overline{m}} - x_1^{\overline{m+1}}\right) + \alpha_{\overline{m}}\left(g^{\overline{m}}\right)^T\left(x_1^* - x_1^{\overline{m}}\right)$.

For any$i \in \left[\overline{m}, \overline{m+1} - 1\right], g^i = g^{\overline{m}} = g^{\overline{m+1}-1}$ and $\widehat{q}^i(x_1) = \widehat{q}^{\overline{m}}(x_1) = \widehat{q}^{\overline{m+1}-1}(x_1)$ for any $x_1 \in X$. Therefore,

$$T^{\overline{m+1}} - T^i \le \alpha_{\overline{m}}\left(g^{\overline{m}}\right)^T\left(x_1^i - x_1^{\overline{m+1}}\right) + \alpha_{\overline{m}}\left(g^{\overline{m}}\right)^T\left(x_1^* - x_1^i\right) \tag{21}$$

For any $i \in \left[\overline{m}, \overline{m+1} - 1\right]$ and $j \in \left[\overline{m+1}, \overline{m+2} - 1\right], \widehat{Q}^j = \widehat{Q}^{\overline{m+1}} = \widehat{Q}^{\overline{m+2}-1}$ for any $x_1 \in X$. Therefore,

$$T^j - T^i \le \alpha_{\overline{m}}\left(g^{\overline{m}}\right)^T\left(x_1^i - x_1^j\right) + \alpha_{\overline{m}}\left(g^{\overline{m}}\right)^T\left(x_1^* - x_1^i\right) \tag{22}$$

□

To the proof of Theorem 1, we here consider two scenarios. For the first scenario, ProSHLA does not stop in a given update process. Thus, any update process exhibits finite iterations before the algorithm stops, which means $\overline{m+1} - \overline{m} < M$ for any m (M represents a large number). For the second scenario, ProSHLA might terminate in a given update process. In the following text, Theorem 1 is proven for each scenario.

Scenario 1: ProSHLA does not stop in a given update process.

In the first scenario, a subsequence of $\{x_1^k\}$, $\{x_1^{\overline{m}}\}$ are considered. We will prove that the subsequence $\{x_1^{\overline{m}}\}$ converges to the true optimal x_1^*. According to the definition of $g^k \in \partial Q(x_1^k, \omega^{k+1})$, we can obtain the following inequality

$$(g^k)^T (x_1^* - x_1^k) \leq Q(x_1^*, \omega^{k+1}) - Q(x_1^k, \omega^{k+1}) \tag{23}$$

where $Q(x_1, \omega^{k+1})$ represents the operational cost function given the outcome ω^{k+1}. According to Lemma 1, we can obtain the following inquality:

$$\alpha_{\overline{m}} (g^{\overline{m}})^T \left(x_1^i - x_1^j\right) \leq (\alpha_{\overline{m}} c_1)^2 / b \tag{24}$$

On the basis of Lemma 3, the difference $T^{\overline{m+1}} - T^{\overline{m}}$ can be described as follows:

$$\begin{aligned} T^{\overline{m+1}} - T^{\overline{m}} &\leq \alpha_{\overline{m}} (g^{\overline{m}})^T \left(x_1^{\overline{m}} - x_1^{\overline{m+1}}\right) + \alpha_{\overline{m}} (g^{\overline{m}})^T (x_1^* - x_1^{\overline{m}}) \\ &\leq -\alpha_{\overline{m}} \left(Q\left(x_1^{\overline{m}}, \omega^{\overline{m+1}}\right) - Q\left(x_1^*, \omega^{\overline{m+1}}\right)\right) + \alpha_{\overline{m}} (g^{\overline{m}})^T (x_1^* - x_1^{\overline{m}}) \\ &\leq -\alpha_{\overline{m}} \left(Q\left(x_1^{\overline{m}}, \omega^{\overline{m+1}}\right) - Q\left(x_1^*, \omega^{\overline{m+1}}\right)\right) + (\alpha_{\overline{m}} c_1)^2 / b \end{aligned} \tag{25}$$

Conditional expectation of Eq. (27) with respect to $\mathcal{H}_k$ can be taken on both side and then we can obtain

$$E\left(T^{\overline{m+1}} | \mathcal{H}_{\overline{m}}\right) \leq T^{\overline{m}} - \alpha_{\overline{m}} (\overline{Q}(x_1^{\overline{m}}) - \overline{Q}(x_1^*)) + (\alpha_{\overline{m}} c_1)^2 / b$$

where $\overline{Q}(x_1)$ represents the expected recourse function, that is $E_\omega Q(x_1, \omega)$. Given the conditioning on $\mathcal{H}_k$, $T^{\overline{m}}$, $\alpha_{\overline{m}}$ and $x_1^{\overline{m}}$ on the right-hand side are deterministic. The conditioning $\mathcal{H}_k$ cannot provide any information on $\omega^{\overline{m+1}}$. Hence, we replace $Q\left(x_1, \omega^{\overline{m+1}}\right)$ (for $x_1 = x_1^k$ and $x_1 = x_1^*$) with its expectation $\overline{Q}(x_1)$. Given that $\alpha_{\overline{m}} (\overline{Q}(x_1^{\overline{m}}) - \overline{Q}(x_1^*)) \geq 0$, the sequence

$$W^{\overline{m}} = T^{\overline{m}} + (\alpha_{\overline{m}} c_1)^2 / b \tag{26}$$

is a positive supermartingale. Theorem 2 implies the almost sure convergence of $W^{\overline{m}}$. Hence,

$$T^{\overline{m}} \to T^* \quad a.s. \tag{27}$$

We perform the summation of Eq. (27) from 0 to $\overline{M}$ and obtain the following inequality:

$$T^{\overline{M}} - T^0 \leq -\sum_{\overline{m}=0}^{\overline{M}} \alpha_{\overline{m}} \left(Q\left(x_1^{\overline{m}}, \omega^{\overline{m+1}}\right) - Q\left(x_1^*, \omega^{\overline{m+1}}\right)\right) + \sum_{\overline{m}=0}^{\overline{M}} (\alpha_{\overline{m}} c_1)^2 / b \tag{28}$$

We take the expectation of both sides. We take the conditional expectation with respect to $\mathcal{H}_{\overline{m}}$ and then over all $\mathcal{H}_{\overline{m}}$ for the first term on the right-hand side.

$$E\left(T^{\overline{M+1}} - T^{0}\right)$$

$$\leq -\sum_{\overline{m}=0}^{\overline{M}} E\left\{E\left\{\alpha_{\overline{m}}\left(Q\left(x_1^{\overline{m}}, \omega^{\overline{m+1}}\right)\right)\right\}\right\} - Q\left(x_1^{*}, \omega^{\overline{m+1}}\right)\right\}\mathcal{H}_{\overline{m}}\right\}\right\} + E\left\{\sum_{\overline{m}=0}^{\overline{M}} \frac{\left(\alpha_{\overline{m}} c_1\right)^2}{b}\right\}$$

$$\leq -\sum_{\overline{m}=0}^{\overline{M}} E\left\{\alpha_{\overline{m}}\left(Q\left(x_1^{\overline{m}}, \omega^{\overline{m+1}}\right) - Q\left(x_1^{*}, \omega^{\overline{m+1}}\right)\right)|\mathcal{H}_{\overline{m}}\right\}$$

$$+ \left(c_1\right)^2/b \sum_{\overline{m}=0}^{\overline{M}} E\left\{\alpha_{\overline{m}}^2\right\}$$

We take the limit as $\overline{M} \to \infty$ and use the finiteness of $T^{\overline{M}}$ and $\sum_{\overline{m}=0}^{\overline{M}} E\left\{\alpha_{\overline{m}}^2\right\}$ to obtain

$$\sum_{\overline{m}=0}^{\overline{M}} E\left\{\alpha_{\overline{m}}\left(Q\left(x_1^{\overline{m}}, \omega^{\overline{m+1}}\right) - Q\left(x_1^{*}, \omega^{\overline{m+1}}\right)\right)|\mathcal{H}_{\overline{m}}\right\} < \infty \tag{29}$$

Given that $Q\left(x_1^{\overline{m}}, \omega^{\overline{m+1}}\right) - Q\left(x_1^{*}, \omega^{\overline{m+1}}\right) \geq 0$ and $\sum_{\overline{m}=0}^{\overline{M}} E\left\{\alpha_{\overline{m}}^2\right\} = \infty (a.s.)$, there exists a subsequence $\{\overline{m}\}$ such that

$$\overline{Q}\left(x_1^{\overline{m}}\right) \to \overline{Q}\left(x_1^{*}\right) \quad a.s.$$

By continuity of $\overline{Q}$, the sequence converges. Hence,

$$x_1^{\overline{m}} \to x_1^{*} \quad a.s.$$

Subsequently, we construct another subsequence $\left\{x_1^{\overline{m}-1}\right\}$. Based on Eq. (27),

$$E\left(T^{\overline{M+2}-1} - T^{\overline{M+1}-1}\right) \leq -\sum_{\overline{m}=0}^{\overline{M}} \alpha_{\overline{m}}\left(Q\left(x_1^{\overline{m+1}-1}, \omega^{\overline{m+2}-1}\right) - Q\left(x_1^{*}, \omega^{\overline{m+2}-1}\right)\right) + \left(\alpha_{\overline{m}} c_1\right)^2/b$$

Like-wise, the following approximation can be proved:

$$x_1^{\overline{m}-1} \to x_1^{*} \quad a.s.$$

By analogic condition, a very general subsequence $\left\{x_1^{i}\right\}, i \in \left[\overline{m}, \overline{m+1} - 1\right]$ will almost surely converge to x_1^{*}. Here, we term this type of subsequence X_s.

In the procedure of ProSHLA, the number of all update iterations is finite. Thus, for any subsequence of $\left\{x_1^{k}\right\}$, we can obtain a subsequence that always belongs to X_s. Then, the following conclusion can be obtained:

$$x_1^k \to x_1^* \quad \text{a.s.}$$

Scenario 2: ProSHLA halts in a given update process.

For the second scenario, ProSHLA halts in a projection procedure which generates a convergent sequence.

Hence, the conclusion of Theorem 1 can be finally obtained. □.

The above analytical processes demonstrate the convergence property of ProSHLA. According to the above results, we require the function $\widehat{Q}(x)$ to be piecewise linear convex. However, for practitioners are interested in a practically scenarios, in which they usually use separable functions to approximate the expected recourse function for stochastic programs with network recourse. Based on Eq. (4), if the separable functions are piecewise linear or purely linear, then practitioners can easily solve this network recourse problem, because a pure network flow problem is polynomial solvable. In the following section, we will discuss this special practice scenario.

3. Application for two-stage stochastic programs with network recourse using separable piecewise linear functions

In this section, we will discuss the scenario where $\widehat{Q}(x)$ is separable for two-stage stochastic programs with network recourse. For this scenario, we can simplify implement ProSHLA without projection step. We denote this simplified version as the Stochastic Hybrid Learning Algorithm (SHLA), which is described as follows (**Figure 2**).

Essentially, SHLA is not convergent. However, if it is applied to two-stage stochastic programs with network recourse, SHLA will enjoys several merits as follows: (1) the solution of $Q(x,\omega)$ is naturally integer; (2) at each iteration, problem $Q(x,\omega)$ is simple network flow problem that can be solved by polynomial algorithm.

Here, if we use separable functions, then assumption (A.6) can be satisfied by the following artificially expression:

$$\widehat{q}^0(x_i) < \widehat{q}^0(x_i + \delta)$$

Note that for both ProSHLA and SHLA, it allows to choose initial approximation function with different value of δ flexibly. Thus, if δ is set to be 1 for any i, then we can guarantee the following expression:

SHLA

Step 1. Set the iteration counter $k = 0$. Construct an initial piecewise linear *convex* function $\hat{Q}^0(x)$. Maintain a sequence of points $\{x^k\}$.

Step 2. Solve the problem $x^k = \text{argmin } \hat{Q}^k(x)$ and obtain $\hat{q}^k(x)$.

Step 3. Obtain g^k. Update $\hat{Q}^k(x)$ by

$$\hat{Q}^{k+1}(x) = \hat{Q}^k(x) + \alpha_k(g^k - \hat{q}^k(x))^T x \qquad (32)$$

Step 4. Check for termination (e.g. an improvement in $\hat{Q}^k(x)$ in the last K iterations). If the check fails, then set $k = k + 1$ and go to Step 2; otherwise, terminate.

Figure 2.
Description of SHLA.

$$\hat{q}_i^0(x_i) < \hat{q}_i^0(x_i + \delta). \tag{A.7}$$

Then, we can reach Theorem 3 below.

Theorem 3. If (A.7) is satisfied, SHLA is always convergent for two-stage stochastic programs problem with network recourse.

Proof. For any $x, y \in X$, if there are unequal, we can obtain the following expression according to (A.7).

$$\hat{q}^k(x) \neq \hat{q}^k(y)$$

Thus,

$$|\hat{q}^k(x) - \hat{q}^k(y)| > 0$$

Hence, if we set $\delta = 1$ and apply ProSHLA for two-stage stochastic programs with network recourse, then ProSHLA can drop the projection step because $\hat{q}^k(x)$ and $\hat{q}^k(y)$ are always unequal. In this situation, ProSHLA is equivalent to SHLA, so SHLA is convergent.

According to the above analysis, we have provided first theoretical convergence support for SHLA-type algorithms which are widely used in numerous applications as mentioned in introduction part. Compared with SHAPE, SHLA does not contain any nonlinear terms so that it can be very efficient. Besides, SHLA can automatically maintain the convexity of the approximation function if the initial piecewise linear functions are properly constructed.

4. Experimental results for performance analysis

In this section, we use two experimental designs to evaluate the performance of the algorithms: (1) An empty container repositioning problem which arises in the context of two-stage stochastic programs with network recourse; and (2) a high dimensional resource allocation problem as an extension experiment. In this section, the empty container repositioning problem is first introduced and then we present the efficiency of ProSHLA and SHLA. Sub-sequentially, we present the convergence of ProSHLA and SHLA, and examine how δ affects convergence performance, and compare the performance under different distributions of random demands. Finally, an extension experiment on a high dimensional resource allocation problem is conducted to evaluate the efficiency our algorithms.

4.1 Problem generator for the empty container repositioning problem

In this subsection, we test our algorithms in an empty container repositioning problem faced by a major Chinese freight forwarder, who need to manage their numerous empty container in a port network which is located in Pearl River Delta in a fixed route from [23]. The port network contains several hubs (large ports) and spokes (small ports). And the demand of empty container is usually uncertain. When the forwarder need to decide the quantity of empty container to ship from one port to another, they did not know the exact demand of container in the future [24, 25]. Thus, we can formulate the problem as a two-stage stochastic programs with network recourse. Before we formally introduce the problem, we present the following notations.

L	=	set of ports;
d_{ij}	=	demand from port i to port j in stage 1;
D_{ij}	=	demand from port i to port j in stage 2;
s_i	=	initial number of empty containers at port i;
S_i	=	number of empty containers at port i in stage 2;
c_{ij}	=	cost for moving an empty container from port i to port j;
r_{ij}	=	profit for moving a laden container from port i to port j;
x_{ij}	=	number of laden containers shipped from port i to port j in stage 1;
y_{ij}	=	number of empty containers shipped from port i to port j in stage 1;

Then, the problem can be formulated as follows:

$$min \sum_{i\in L}\sum_{j\in L}\left\{-r_{ij}x_{ij}+c_{ij}y_{ij}\right\}+E_{\omega}[Q(x,\omega)], \tag{30}$$

s.t.,

$$\sum_{j\in L}\left\{x_{ij}+y_{ij}\right\}=s_i, \quad \forall i\in L \tag{31}$$

$$\sum_{i\in L}\left\{x_{ij}+y_{ij}\right\}=s_j, \quad \forall j\in L \tag{32}$$

$$y_{ij}\geq 0, \ \forall i,j\in L \tag{33}$$

where the recourse function $Q(x,\omega)$ is given as follows:

$$Q(x,\omega)=min \sum_{i\in L}\sum_{j\in L}\left\{-r_{ij}x_{ij}(\omega)+c_{ij}y_{ij}(\omega)\right\} \tag{34}$$

s.t.,

$$\sum_{j\in L}\left\{x_{ij}(\omega)+y_{ij}(\omega)\right\}=s_i, \quad \forall i\in L \tag{35}$$

$$\sum_{i\in L}\left\{x_{ij}(\omega)+y_{ij}(\omega)\right\}=s_j, \quad \forall j\in L \tag{36}$$

$$y_{ij}(\omega)\geq 0, \ \forall i,j\in L \tag{37}$$

In order to evaluate the algorithm, a set of problem instances are created. In this study, the problem generator creates ports in L in a 100-mile by 100-mile rectangle. We simply use the Euclidean distance between each pair of ports as the corresponding travel distance. We set the holding cost for a demand to 15 cents per time instance. We set the net profit for a demand to 500 cents per mile. The empty cost is set to 40 cents. The demand D_{ij} between locations i and j is set as follows:

$$D_{ij} = out_j \cdot in_i \cdot v,$$

where
out_j = outbound potential for port j;
in_i = inbound potential for port i;
v = random variable.

The outbound and inbound potentials for each port represent the capability of the location to generate outbound demand or attract inbound containers. In the generator, We draw the inbound potential, in_i, for port i between 0.2 and 1.8 uniformly, while the corresponding out_j is set as $out_j = 2 - in_i$. The reason for this setting is that in real-world regions, large inbound flows port usually exhibits small outbound flows. We also include a random number v with mean 30, that is, the typical daily demand between each pair of locations to capture the randomness in demand. In order to test the performance of the algorithms under different distributions, we also evaluate the performance under exponential, normal and uniform distribution. We set the stepsize α_k to $1/k$.

We solve a deterministic network flow problem to construct an initial piecewise linear functions as described in [1], and we replace the random demand by their mean values in the deterministic problem. Then, we can obtain $\overline{S} = \{\overline{S}_1, \overline{S}_2, \dots, \overline{S}_n\}$. For each $i \in L$, we generate the initial approximation function $\widehat{Q}_i^0(x) = c(x - \overline{S}_i)^2, x = 0, \delta, \dots, k\delta, \dots .K\delta$, where c is a positive parameter and $x \in [0, K\delta]$. In the projection step, a least-squares problem is solved as following:

$$x^{k+1} = \text{argmin}\left(x^{k+1} - \left(x^k + \alpha_k g^k\left(x^k\right)\right)\right)^2, x^{k+1} \in X.$$

4.2 Effectiveness and efficiency performance

To test the efficiency of the algorithm, we use a myopic algorithm, a posterior bound (PB), the L-shaped algorithm [12] and the inexact cut algorithm [15] as benchmarks. The myopic algorithm simply solves a static deterministic assignment problem at the current stage while ignoring uncertainties in the second stage. It is necessary to solve a deterministic network flow problem with all realized demands to obtain PB. Note that such a posterior optimization involves no uncertainty since decisions are allowed to anticipate future demand. Thus, the cost of PB is the lowest and normally unreachable. As for the L-shaped algorithm and the inexact cut algorithm, a group of linear programming problems with valid cuts should be solved.

We use 8 instances, in which the number of empty containers is ranged from 400 to 3200, and the corresponding number of ports is ranged from 5 to 40. For each instance, 2000 samples are implemented and we obtain the solutions of the myopic algorithm and the sample means of PB, the inexact cut algorithm, the L-shaped algorithm, SHLA and ProSHLA. For SHLA, two classes of initial functions with δ = 1 and δ = 2 are selected, whereas we select δ = 2 for ProSHLA.

We show the experiment results on total cost in **Table 1**. In **Table 1**, column 1 presents the number of ports, and column 2 shows the number of the empty containers. The PB bounds are contained in column 3. Columns 4–9 contain the solutions achieved by the myopic algorithm, the L-shaped algorithm, the inexact cut algorithm, SHLA-1, SHLA-2 and ProSHLA, respectively. From the table, it clearly demonstrates that the inexact cut algorithm, the L-shaped algorithm, SHLA, and ProSHLA can achieve optimal or very-near-optimal solutions, which are closer to the PB (lowest)

N_L	N_R	Total cost (dollars)						
		PB	Myopic	L-shaped	Inexact cut	SHLA-1	SHLA-2	ProSHLA
5	400	−28,551,302	−27,761,331	−28,551,274	−28,551,227	−28,464,248	−28,463,941	−28,464,002
10	800	−59,397,423	−58,432,159	−59,396,702	−59,396,319	−59,338,653	−59,338,190	−59,338,341
15	1200	−98,451,193	−93,188,576	−98,449,868	−98,449,427	−98,257,484	−98,257,244	−98,257,395
20	1600	−147,390,005	−141,062,187	−147,388,360	−147,387,532	−147,269,239	−147,269,212	−147,269,213
25	2000	−185,223,883	−180,875,614	−185,220,423	−185,219,663	−185,092,289	−185,092,113	−185,092,143
30	2400	−234,005,740	−226,978,090	−234,004,427	−234,003,513	−233,401,849	−233,401,516	−233,401,658
35	2800	−266,375,728	−260,966,356	−266,375,468	−266,374,497	−266,337,862	−266,337,649	−266,337,660
40	3200	−304,910,355	−293,882,881	−304,908,597	−304,906,829	−304,907,153	−304,906,829	−304,906,962

Table 1.
Total cost for SHLA and ProSHLA.

N_R	N_L	Computation time (s)				
		Inexact cut	**L-shaped**	**ProSHLA**	**SHLA-1**	**SHLA-2**
400	5	76	153	43	28	28
800	10	382	535	148	90	95
1200	15	598	1140	332	224	217
1600	20	1084	2277	628	417	420
2000	25	1843	4190	1107	676	790
2400	30	2338	5491	1559	1112	1066
2800	35	4249	8075	2341	1539	1531
3200	40	8154	18,636	5307	3010	3428

Table 2.
Computational time of ProSHLA and SHLA.

bounds than those of the myopic method. Moreover, the solutions of the L-shaped algorithm are the best solutions known, which are slightly better than the inexact cut algorithm because the latter produces valid cuts that are inexact in the sense that they are not as constraining as optimality cuts in the Lshaped algorithm. In addition, the performance of SHLA ($\delta = 1$) outperforms that of SHLA ($\delta = 2$) and ProSHLA ($\delta = 2$), the reason is that small δ can lead to good performance. A specific discussion with impact of δ will be demonstrated later. The performance of ProSHLA ($\delta = 2$) is slight better than SHLA ($\delta = 2$). Because the projection steps in ProSHLA help improve the solution. Considering the speed of convergence is quite important in practical problems, we will focus on the computational time for different algorithms, which is shown in the **Table 2** below.

As shown in **Table 2**, ProSHLA and SHLA are more efficient than the inexact cut algorithm and the L-shaped algorithm because ProSHLA and SHLA can utilize the network structure while using the stochastic sub-gradient to approximate the recourse function. From the table, we find that the inexact cut algorithm and L-shaped algorithm are time-consuming, the reason is that here are 2000 samples, and it corresponds to a very large number of cuts for the inexact cut algorithm and L-shaped algorithm. It can also be observed that the computational time of the inexact cut algorithm is smaller than that of the L-shaped algorithm, and the reason is that the optimality cut in L-shaped is more than the valid cuts in the inexact cut algorithm. Moreover, the computational time of SHLA ($\delta = 1$) is almost equal to that of ($\delta = 2$), which reveals that the computational time canot be affect by the choice of δ. In contrary, ProSHLA($\delta = 2$) requires more computational time than SHLA ($\delta = 2$), and the reason is that the projection step in ProSHLA are time-consuming. In the following text, we focus on the convergence performance of SHLA and ProSHLA. Thus, only the results of the myopic algorithm, PB, ProSHLA and SHLA are demonstrated, and we use the solutions of the myopic algorithm and PB as the upper and lower bounds, respectively.

4.3 Analysis of convergence performance

In this subsection, a set of experiments are conducted to evaluate the convergence performance of SHLA and ProSHLA, and we choose the second instance ($N_R = 800$

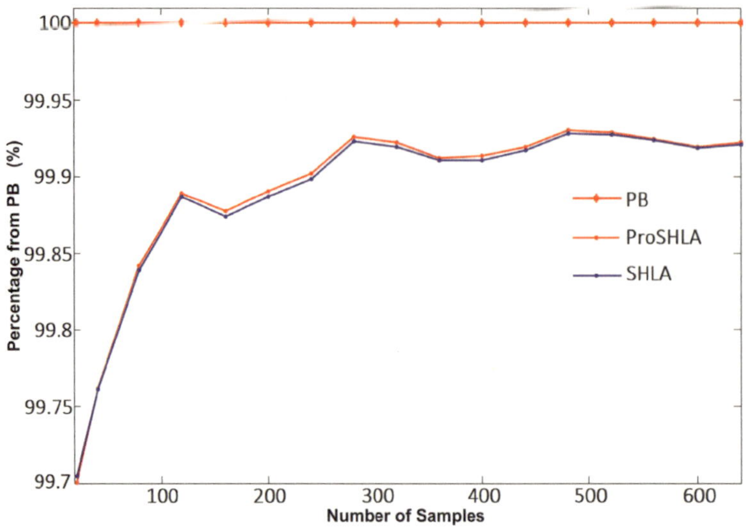

Figure 3.
Convergence rate of ProSHLA and SHLA.

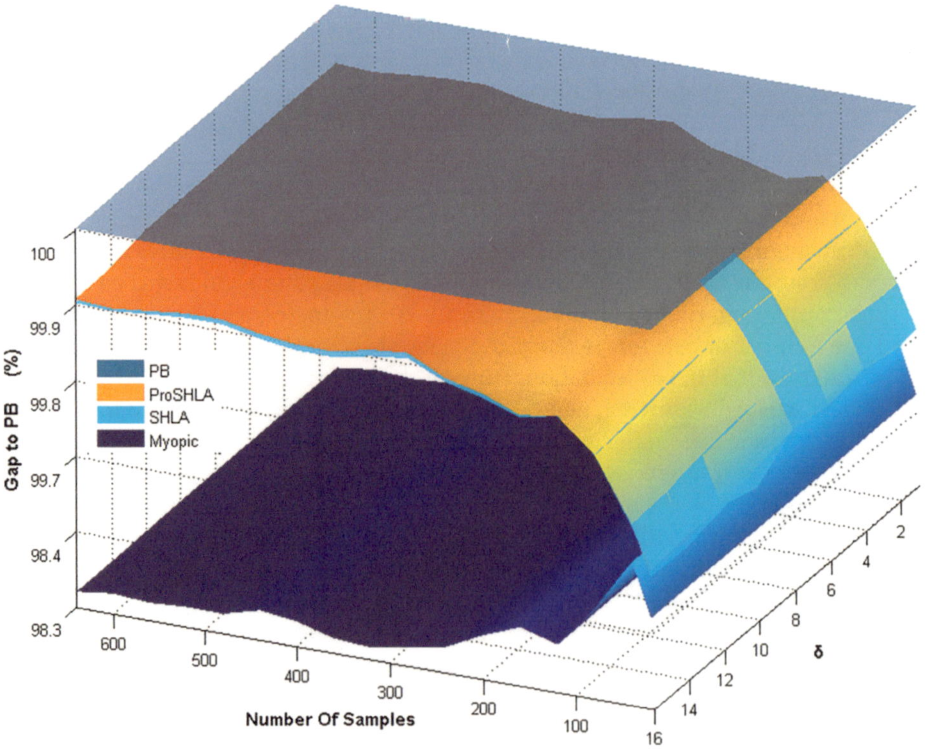

Figure 4.
Gaps to PB for various δ.

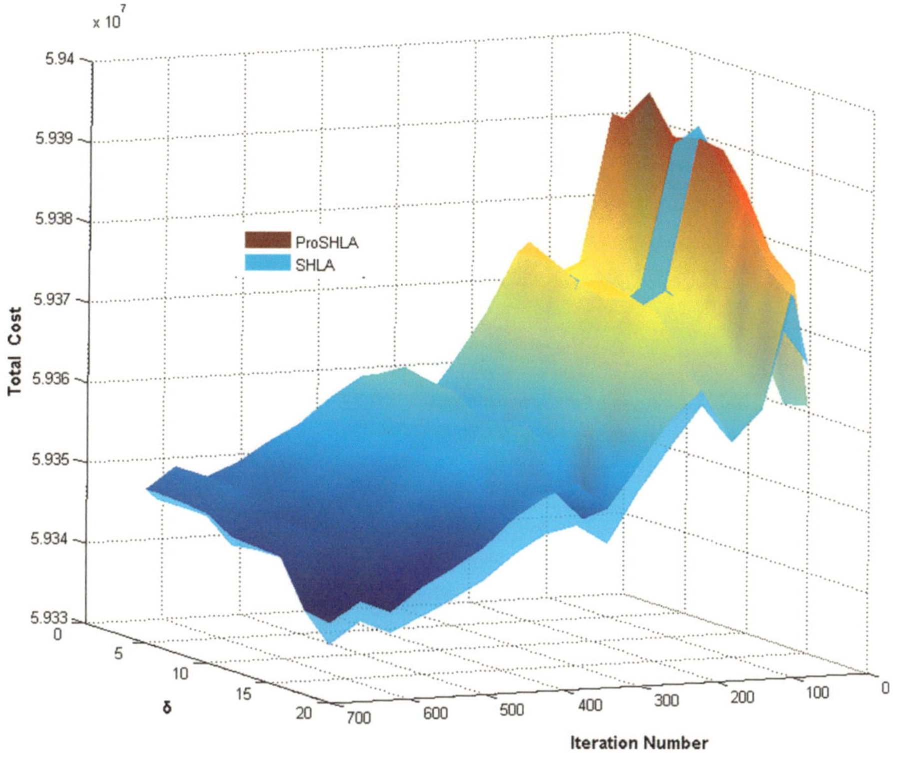

Figure 5.
Comparison of ProSHLA and SHLA for various δ.

and N_L = 10) as the experimental illustration. The range of the sample number is set from 20 to 640, and we record the result of each combination of N_R and N_L at each iteration. We can seem from **Figure 3** that the convergence rate of SHLA-1 and ProSHLA is remarkably high.

To further evaluate how δ affects the algorithm's convergence performance, a set of computational experiments are conducted. We increase δ from 1 to 16 and the number of samples from 20 to 640. Here are many combination of δ and the number of samples. We record the sample means of the solutions of SHLA and ProSHLA, PB and the myopic method for each combination. We demonstrate the 3D plots of the solution in **Figures 4** and 5. As in **Figure 4**, the layers of ProSHLA and SHLA are extremely close to the PB layer, and this implies the ProSHLA and SHLA are convergent rapidly for various δ. Furthermore, it can been seem that the performance of ProSHLA can slightly exceeds that of SHLA. In order to further investigate the difference between SHLA and ProSHLA, we demonstrate the performance of ProSHLA and SHLA separately in **Figure 5** (without the myopic algorithm and PB). As described in **Figure 5**, the choice of δ can affect the performance of ProSHLA and SHLA, and a small δ usually leads to a good solution.

We provides more details on the convergence performance of ProSHLA and SHLA for various δ in **Table 3** below, which clearly demonstrates that in conjunction with the small δ, the performance of SHLA and ProSHLA is close to that of PB.

Total cost—% gap to PB (computational time in seconds)				
δ	PB	Myopic	SHLA	ProSHLA
1	−59,397,423	1.6251%	0.0989%(909.5)	0.0989% (1441.5)
2	−59,397,423	1.6251%	0.0997%(907.9)	0.0995% (1506.5)
4	−59,397,423	1.6251%	0.1001%(901.8)	0.0997% (1503.6)
6	−59,397,423	1.6251%	0.1005%(911.5)	0.1004% (1553.5)
8	−59,397,423	1.6251%	0.1028%(901.7)	0.1024% (1535.6)
16	−59,397,423	1.6251%	0.1189%(920.3)	0.1150% (1565.4)

Table 3.
Performance under various δ (no. of samples is 2000).

4.4 An extension experiment on a high dimensional resource allocation problem

Due to the limitation of the container setting, an extension experiment on a higher dimensional problem is considered in this subsection. In this problem, there exists several retailers R and many production facilities (with warehouse) L. In stage 1, an amount x_{ij} is moved to a warehouse or retailer or location j from production facility i before the retail demand is realized. When we know the consumer's demand, then y_{ij} products are moved to retailer location j from production facility i. Besides, the type of the consumer's demand at each location j is different, we denote the type as $t \in T$, we set the consumer's demand of type t at location j as D_j^t, and provide p_i^t unit of type t at production location i. We denote the production capacity of location i by cap_i. This problem is a non-separable problem.

Subsequently, we formulate the problem as follows:

$$min \sum_{i \in L} \sum_{j \in L \cup R} c_{ij}^1 y_{ij} + E_\omega[Q(x,\omega)] \tag{38}$$

subject to

$$\sum_{j \in L \cup R} x_{ij} \le cap_i, \quad \forall i \in L \tag{39}$$

$$\sum_{i \in L} x_{ij} = s_j, \quad \forall j \in L \cup R \tag{40}$$

$$x_{ij}, s_j \ge 0, \quad \forall i \in L, \forall j \in L \cup R \tag{41}$$

where the recourse function $Q(x,\omega)$ is given as follows:

$$Q(x,\omega) = min \sum_{i \in L \cup R} \sum_{j \in R} c_{ij}^2 y_{ij} - \sum_{i \in R} \sum_{t \in T} r_{iL}^t p_i^t \tag{42}$$

subject to

$$\sum_{j \in R} y_{ij} = s_i, \quad \forall i \in L \cup R \tag{43}$$

$$\sum_{i \in L \cup R} y_{ij} = \sum_{t \in T} p_j^t, \ \ \forall i \in R \tag{44}$$

$$p_j^t \leq D_j^t(\omega), \ \ \forall t \in L \cup R, \forall j \in R, t \in T \tag{45}$$

In the first stage, we set $c_{ij}^1 = c_0^1 + c_1^1 d_{ij}$, where d_{ij} is the Euclidean distance between locations i and j, and c_0^1 is the production cost for each product and c_1^1 is the transportation cost per mile. For the second stage costs, we set

$$c_{ij}^2 = \begin{cases} c_1^2 d_{ij} & \text{if } i \in L \text{ or } i = j \\ c_0^2 + c_1^2 d_{ij} & \text{if } i \in R \text{ and } i \neq j \end{cases}$$

c_1^2 is the transportation cost per mile in the second stage, and c_0^2 represents the fixed charge for moving each product from one retailer location to another retailer location. For one unit of the demand type t occurring in retailer location i, a revenue r_i^t will be obtained. Our problem instances differ in the number of products and $|L \cup R|$, and it determines the dimensionality of the recourse function.

Similarly, we use the inexact cut algorithm [15] and the L-shaped algorithm [12] as benchmarks, and these two algorithms are Benders decomposition based methods. Considering the convergence rate is quite important practically, in this part, our main focus is on the speed of convergence. In order to evaluate the speed of convergence of different methods, each algorithm is implemented for 40, 160, 640, 1200, and 4000 iterations, and a side by side comparison of the algorithms has been made when the number of iteration increases. For the L-shaped and inexact cut algorithms, the number of iterations refer to the number of cuts used to approximate the expected recourse function. For ProSHLA (δ = 2), the number of iterations refer to the number of demand samples used.

Table 4 below shows the experiment results. In the experiment, the L-shaped algorithm has been used to help find the optimal solution. In the table, the numbers denote the percent deviation between the optimal value and the objective value.

For all problem instances, we use the L-shaped algorithm to find the optimal solution. The numbers in the table represent the percent deviation between the objective value and the optimal value obtained after a certain number of iterations. The computational time per iteration are also listed in **Table 4**. The computational results on 5 scale of dimensionality instances.

In **Table 4** above, column 1 presents the number of the locations, and column 2 shows the number of the products. Column 3 presnets the method that we used in the experiment. The percent deviation from the optimal value are contained in columns 4 to 8. Column 9 lists the computational time per iteration. According to results in **Table 4**, ProSHLA is able to obtain high quality solutions very efficient for different problem instance, and it can maintain the consistent performance in problem of different sizes, especially for large problems. This performance characteristic makes ProSHLA promising for large-scale application. In comparison with these two Benders decomposition-based methods, ProSHLA is competitive for high dimensional problems. The reason is that separable approximations usually scale much more easily to very high dimensional problems. Note that in the first problem instance, when the number of location is 6 and the number of resource is 10 (the inventory in a location might be 0, 1, 2), the result of ProSHLA seems to be breakdown, because the problem instance in this subsection is non-separable, which may introduce errors when we use

			Number of iterations					
$N_{\|L \cup R\|}$	N_P	***Algorithm***	**40**	**160**	**640**	**1200**	**4000**	**Sec./iter.**
$\|L \cup R\| = 6$	10	ProSHLA	13.26	8.61	2.93	2.92	2.73	0.01
		L-shaped	1.17	0	0	0	0	0.07
		Inexact cut	0.96	0	0	0	0	0.05
$\|L \cup R\| = 10$	200	ProSHLA	10.58	3.01	0.61	0.29	0.12	0.07
		L-shaped	1.85	0	0	0	0	0.26
		Inexact cut	1.31	0	0	0	0	0.21
$\|L \cup R\| = 20$	400	ProSHLA	7.22	1.22	0.42	0.23	0.05	0.30
		L-shaped	10.46	1.16	0	0	0	1.13
		Inexact cut	6.63	0.98	0	0	0	1.04
$\|L \cup R\| = 40$	800	ProSHLA	6.03	0.82	0.34	0.17	0.02	0.83
		L-shaped	23.57	3.23	0.31	0.02	0	9.53
		Inexact cut	17.16	2.24	0.13	0.01	0	8.72
$\|L \cup R\| = 100^*$	2000	ProSHLA	5.68	0.78	0.15	0.04	0 0.03	2.68
		L-shaped	44.84	14.51	1.38	0.5		36.53
		Inexact cut	29.56	8.14	0.91	0.25	0.03	30.98

Note. Figures represent the deviation from the best objective value known.
**Optimal solution not found.*

Table 4.
Percent error over optimal solution with different algorithms costs.

the separable approximations to approximate the expected recourse function. However, it will not happen on large problems. As for large problems, the separable approximations are nearly continuous, rather than being just piecewise continuous.

According to the above computational results, ProSHLA is a promising method for two-stage stochastic programs, but more comprehensive numerical work is needed before using it in a particular problem. Owing to its efficient performance and simplicity, ProSHLA is a very promising candidate for high-dimensional problems. Moreover, we can use it as an initialization routine method for high-dimensional stochastic programming problems, and it can exploit high-quality initial feasible solution.

5. Conclusion

In this study, we propose an efficient machine learning algorithm for two-stage stochastic programs. This machine learning algorithm is termed as projected stochastic hybrid learning algorithm, and consists of stochastic sub-gradient and piecewise linear approximation methods. We use the stochastic sub-gradient and sample information to update the piecewise linear approximation on the objective function. Then we introduce a projection step, which implemented the sub-gradient methods, to jump out from a local optimum, so that we can achieve a global optimum. By the innovative projection step, we show the convergent property of the algorithm for general two-stage stochastic programs. Furthermore, for the network recourse problem, our algorithm can drop the projection steps, but still maintains the convergence property. The computational results reveal the efficiency of the proposed algorithms and the proposed algorithms are distribution-free. Furthermore, the convergence rate can be affected by the granularity of the initial function (δ). Small granularity usually leads to a high convergence rate. Finally, the computational results also show that the proposed algorithm is very competitive for high dimensional problems. Compared with MAT, the proposed algorithm can collect the information based on knowledge gradients and use it to update the recourse function by learning steps. It can overcome the "curse of dimensionality". Moreover, it can transfer the problem into a polynomial solvable problem.

Acknowledgements

This research is partially supported by the National Nature Science Foundation of China (Project No. 71701221) and the Natural Science Foundation of Guangdong Province, China, under Grant number: 2019A1515011127.

Author details

Zhou Shaorui*, Cai Ming and Zhuo Xiaopo
Sun Yat-sen University, Guangzhou, China

*Address all correspondence to: zshaorui@gmail.com

References

[1] Cheung RK, Chen CY. A two-stage stochastic network model and solution methods for the dynamic empty container allocation problem. Transportation Science. 1998;**32**(2): 142-162

[2] Bouzaiene-Ayari B, Cheng C, Das S, Fiorillo R, Powell WB. From single commodity to multiattribute models for locomotive optimization: A comparison of optimal integer programming and approximate dynamic programming. Transportation Science. 2016;**50**:366-389

[3] Moreno A, Alem D, Ferreira D, Clark A. An effective two-stage stochastic multi-trip location-transportation model with social concerns in relief supply chains. European Journal of Operational Research. 2018;**269**(3):1050-1071

[4] Kim K, Mehrotra S. A two-stage stochastic integer programming approach to integrated staffing and scheduling with application to nurse management. Operations Research. 2015;**63**:1431-1451

[5] Wallace SW, Ziemba WT. Applications of stochastic programming. In: MOS-SIAM Series on Optimization. Vol. 5. Philadelphia, PA: Society for Industrial and Applied Mathematics (SIAM); Mathematical Programming Society (MPS); 2005. ISBN:0-8971-555-5

[6] Kleywegt AJ, Shapiro A. Homem de Mello T. the sample average approximation method for stochastic discrete optimization. SIAM Journal on Optimization. 2001;**12**(2):479-502

[7] Ermoliev Y. Stochastic quasigradient methods. In: Numerical Techniques for Stochastic Optimization. New York: Springer-Verlag; 1988

[8] Robbins H, Monro S. A stochastic approximation method. The Annals of Mathematical Statistics. 1951;**22**(3): 400-407

[9] Rockafellar RT, Wets JB. A note about projections in the implementation of stochastic quasigradient methods. In: Numerical Techniques for Stochastic Optimization, Springer Ser. Comput. Math. Vol. 10. Berlin: Springer; 1988. pp. 385-392

[10] Ruszczyñski A. A linearization method for nonsmooth stochastic optimization problems. Mathematics of Operations Research. 1987;**12**:32-49

[11] Benders JF. Partitioning procedures for solving mixed-variables programming problems. Numerische Mathematik. 1962;**4**(1):238-252

[12] Van Slyke RM, Wets RJ-B. L-shaped linear programs with applications to optimal control and stochastic programming. SIAM Journal on Applied Mathematics. 1969;**17**(4):638-663

[13] Pereira MVF, Pinto LMVG. Multi-stage stochastic optimization applied to energy planning. Mathematical Programming. 1991;**52**:359-375

[14] Zakeri G, Philpott AB, Ryan DM. Inexact cuts in benders decomposition. SIAM Journal on Optimization. 2000; **10**(4):643-657

[15] Shapiro A. Analysis of stochastic dual dynamic programming method. European Journal of Operational Research. 2011;**209**(1):63-72

[16] Rebennack S. Combining sampling-based and scenario-based nested benders decomposition methods: Application to stochastic dual dynamic programming.

Mathematical Programming. 2016; **156**(1):343-389

[17] Philpott AB, Guan Z. On the convergence of stochastic dual dynamic programming and related methods. Operations Research Letters. 2008;**36**: 450-455

[18] Girardeau P, Leclere V, Philpott AB. On the convergence of decomposition methods for multistage stochastic convex programs. Mathematics of Operations Research. 2015;**40**(1): 130-145

[19] Cheung RK, Powell WB. SHAPE—A stochastic hybrid approximation procedure for two-stage stochastic programs. Operations Research. 2000; **48**(1):73-79

[20] Powell WB, Ruszczyñski A, Togaloglu H. Learning algorithms for separable approximation of discrete stochastic optimization problems. Mathematics of Operations Research. 2004;**29**(4):814-836

[21] Godfrey GA, Powell WB. An adaptive dynamic programming algorithm for dynamic fleet management I: Single period travel times. Transportation Science. 2002;**36**(1):21-39

[22] Neveu J. Discrete Parameter Martingales. Amsterdam: North Holland; 1975

[23] Song DP, Dong JX. Empty container management in cyclic shipping routes. Maritime Economics & Logistics. 2008; **10**(4):335-361

[24] Zhou S, Zhang H, Shi N, Xu Z, Wang F. A new convergent hybrid learning algorithm for two-stage stochastic programs. European Journal of Operational Research. 2020;**283**(1): 33-46

[25] Xu L, Zou Z, Zhou S. The influence of COVID-19 epidemic on BDI volatility: An evidence from GARCH-MIDAS model. Ocean Coastal Management. 2022. DOI: 10.1016/j.ocecoaman. 2022.106330